T0243254

PERIOD

PERIOD

The Real Story of Menstruation

KATE CLANCY

PRINCETON UNIVERSITY PRESS

PRINCETON & OXFORD

Published by Princeton University Press
41 William Street, Princeton, New Jersey 08540
99 Banbury Road, Oxford OX2 6JX

press.princeton.edu

Library of Congress Control Number: 2022944663

First paperback edition, with discussion questions, 2024
Paper ISBN 9780691264592
Cloth ISBN 9780691191317
ISBN (e-book) 9780691246826

British Library Cataloging-in-Publication Data is available

Editorial: Alison Kalett, Hallie Schaeffer
Jacket/Cover Design: Karl Spurzem
Production: Jacqueline Poirier, Lauren Reese
Publicity: Julie Haav, Kate Farquhar-Thomson
Copyeditor: Wendy Lawrence

This book has been composed in Arno Pro, Balboa, Etna X Condensed

Printed in the United States of America

CONTENTS

PREFACE

THERE WERE a few key things I learned about periods as an adolescent. From fifth-grade health class, I learned that menstruation signaled a failed menstrual cycle with no baby. From my pediatric nurse practitioner Dr. Debbie, I learned that periods make you iron deficient. And from the world around me, I learned that I must hide all signs that I menstruated or face deep, crushing shame. This last lesson led to my acquiring the ability to plaster a smile on my face while passing a clot and developing an obsessive habit of checking to make sure no menstrual blood is visible anywhere on my body.

That warning from Dr. Debbie imprinted itself on me as well, to the point that as a young adult I went down a rabbit hole reading as much as I could about iron and menstruation. It even motivated a small project I undertook as part of my dissertation in graduate school. Using samples from my fieldwork in rural Poland, I performed an analysis comparing my research participants' endometrial thickness (an ultrasound measurement of how much endometrium, or uterine lining, you have) with their red blood cell and hemoglobin levels (both indicators of iron status). What I found was the opposite of what was medically assumed: the thicker your endometrium, the *better* your iron status.[1]

This would be the first of many moments when I noticed an assumption about periods that, with even the gentlest of prodding, completely disassembled. And I could not help but also notice that underlying these assumptions was a certain belief about what people, organs, or processes carry scientific importance, especially in my discipline of anthropology. The origin stories in my field include such winning narratives as "White people are the pinnacle of the human race," and

"Men were the hunters and their behavior drove all of human evolution." There were anthropologists already contesting these narratives by the time I was in college and graduate school, but even thoughtful critiques were met with contempt by many whom I read in the literature and met at conferences. I had one professor who only assigned women anthropologists in his one "feminist" week of the semester, but we had to read these works alongside scornful critiques. In my reflection assignment that week, I wrote that it seemed like he was setting up these authors to be mocked. In response, he read my comments aloud to the class and laughed. That moment created in me my own personal spite project to prove a different type of science is possible—that someone like me, asking the questions I ask, could be a professor. A PhD and a stint teaching college composition later, and I became a professor through a combination of fortune and privilege. I've been studying menstrual cycles off and on ever since.

When I decided I wanted to write this book, I was years deep into a different spite project that had grown to epic proportions: someone dared to tell me that uncovering discrimination in the sciences was a "witch hunt." It started with a collaboration to study sexual harassment in the field sciences, which led to additional projects in astronomy and the planetary sciences, then undergraduate physics, and then a major consensus report, testifying in front of Congress and flying all over the country for a year disseminating the results of the report.[2]

As a survivor of sexual trauma multiple times over, the work I was involved with was challenging. I was fueled by it, but I also hated it. I developed vicarious trauma from all of the stories of assault, harassment, and discrimination I absorbed as both researcher and advocate, and it seeped into and mixed with my own experiences. I started heavily compartmentalizing my work and home life. While there were many wins, there were too many losses: most frustrating was realizing the changes institutions needed to make were not the ones they wanted to hear about. They wanted to sponsor a few talks, add online training. I wanted them to overhaul the incentive structures that permit bad behavior and stop thinking so short term with their risk management.

After the birth of my second child, I struggled to see any physical touch as kindly intended (except from my infant, barely). Men were looming and frightening, even my own partner. Therapy got me clear enough on things to realize I needed to extricate myself from sexual harassment research and advocacy work, at least for a time. All I wanted to do was get back to the whole reason I became an anthropologist and study periods again.

The timing of the creation of this book and my sabbatical the year after my second child was born saved my life. I wanted to be a nerd again. I wanted to read about spiral arteries. I wanted to look at ultrasounds. Like so many of the sexual harassment survivors I've interviewed over the years, I just wanted to get back to my science. This project allowed me to go back to being the academic who eagerly shares how tissue remodeling works in the uterus to friends, family, and strangers. I started a podcast as an excuse to interview all sorts of dream guests who could teach me even more.

At the same time, in getting back to menstruation science I also found I couldn't write a straight science book. It was impossible, not just because in writing about menstruation I found I could not escape the eugenic history of anthropology and gynecology or even how that history reverberates and affects reproductive justice in the present. It was because there are far more interesting questions to ask, wondrous and curious and weird things to explore, when we are able to step away from how the historical study of menstruation and menstrual cycles has shaped our knowledge. There are waves of competition between ovarian follicles.[3] There are muscle contractions that ripple up and down the uterus.[4] There are crypts in the cervix that store sperm for when the uterus feels like using them.[5] There are chemicals in menstrual fluid that help repair tissue.[6]

Menstruation is a wild process that should captivate and delight. It offers up so many lessons in terms of how we understand bodily autonomy, sexual selection, even tissue engineering. It is strange, then, that instead of being something so fundamental to science education as Mendel's peas or dinosaur bones or the planets of our solar system, it gets at best a brief mention in health class. Because of the history of how knowledge is produced, disseminated, and used in dominant culture, two things have happened instead. Within science and especially medical structures, knowledge is power and therefore often withheld; within political structures correct knowledge is rarely the point. Fertility control has been on the agenda in gynecology and in politics for decades, if not centuries. So we know little about menstruation, and what's worse, what we often know is wrong.

On the one hand, much of the reason why what we know about menstruation is wrong—why we characterize it as burdensome or disgusting—comes from legacies of colonialism, racism, and sexism that pervade science, as I will soon show. Maybe by the end of this book, you might feel a little less revolted, a little more friendly, about periods. On the other hand, that does not mean the goal of this book is to show you the histories and subjectivities of menstruation science so you will make any different decisions about your body. With this book I want you to be able to cast your eyes upward, at the systems that persist in putting limits on better understanding, caring for, and treating bodies who menstruate. There will be marvels; there will be pointed critiques. There will not be instructions for IUD self-removal.

INTRODUCTION

Taking the Mystery Out of
Menstruation

IN WRITING this book, I was confronted with a lot of disgust. Don't get me wrong—plenty of people got it and cheered me on. But I still encountered a number of people who did not get why I wanted to write about menstruation. Yet addressing the cultural prevalence of menstrual disgust, stigma, and avoidance is important to recognize why menstruation is understudied and often misunderstood.

My favorite-slash-least-favorite example of this phenomenon comes from a study by Dr. Tomi-Ann Roberts, a professor of psychology at Colorado College.[1] Roberts hired women actors to pretend to be fellow participants in a study on "group productivity." The actor would join the participant in a room, and the two would sit and fill out one packet of questionnaires together. While the researcher left to get the second packet of activities, the actor would pretend to fumble in her handbag for some lip balm. Under one condition, she would drop a tampon in front of the participant, and in the other condition, she would drop a hair clip. She would then return the tampon or hair clip to her handbag, get out the lip balm, use it, and put it back. The researcher would come back, have both participant and actor fill out a second packet of forms with key questionnaires actually needed for the research, and then lead them to a waiting area. The actor would sit in one chair at the end of a row in order to leave the participant a range of chairs to choose from, closer to or farther from her.

What Roberts and colleagues found using that second stack of questionnaires is that the participants who saw the actor drop a tampon viewed her as less competent and liked her less. These findings were true regardless of the research participant's gender. The participants exposed to the tampon-dropping condition were also then more likely to objectify women compared to those exposed to a hair clip drop. And the waiting area experiment, where the research participant could choose to sit closer to or farther from the actor? Fifty three percent of the participants with the tampon-dropping actor sat farther away, compared to only 32 percent of those hanging out with the hair clip–dropping actor, suggesting a disgust reaction among those exposed to the tampon.

Negative attitudes about menstruation are pervasive, and one of the main costs is silence. In the United States, for example, these attitudes fall along three axes: those of concealment, activity, and communication. American girls are taught by parents, teachers, and the broader public to conceal when they are menstruating and to reduce physical activity. The communication taboo often limits their ability to gain information about the practice, management, and experience of periods.[2] Similarly, focus group research on British girls found that they conceive of menstruation as shameful, as something to be hidden, and as a form of illness.[3] Most menstruating people when surveyed regret that young people are taught so little about what periods will actually feel like and what to do once you start getting them.[4] In a culture that still inherently sees menstruation as shameful and dirty, what gets neglected is engaging directly with menstrual management, with experiences of menstrual bleeding and with blood, in order to help children figure out what periods are like and how to manage them day-to-day.

Damaging views toward menstruation extend beyond menstruating people themselves. A study in Taiwan found that, despite education programs on menstruation at school, the boys in the sample had a significantly worse attitude toward menstruation than the girls.[5] An older study from the United States showed that men tended to think the majority of menstrual symptoms occurred during the menstrual phase, whereas women reported that they occurred during the premenstrual phase. Men also tended to think periods were more emotionally debilitating but less physically bothersome than women.[6] Hostile beliefs about menstruation can even lead to stereotype threat: giving women a menstrual distress questionnaire before performing other cognitive tasks decreases their task performance.[7] These beliefs get ground in.

These beliefs are in me too. I was surprised to notice that when my oldest wanted to use a tampon for the first time, I had to talk myself into slowing down, explaining everything fully, and being in the bathroom with them to coach them through the experience. Initially, I had wanted to just hand them a tampon, give them a quick rundown, and be done with it. It was only because my kid was bright enough and brave enough to keep asking me questions—but what part goes in? Does the whole thing stay in? What do I use to get the tampon applicator up in my vagina? How will I know if I've done it right?—that I realized my aversive behavior and corrected it.

Despite the sweeping nature of menstrual stigma, menstrual cycles do get studied in my field, anthropology, because of their relevance to reproduction and therefore to natural selection and evolution (since it is pretty concerned with how effective a given organism is at passing on its genes to the next generation). A lot of what we study concerns variation: how the timing of first periods and last periods, the concentrations of hormones, and the thickness of the endometrium vary with lifestyle and environment. Something I return to again and again in this book is how the physiology of the uterus is defined by its own flexibility in the face of stressors from the environment. Sometimes ovarian hormones are suppressed, or menstrual cycles even stopped, as a way for the body to adaptively respond to poorer conditions. This is not the same thing as free will or choice. But it ties in with a major element of human evolution—that humans have always made choices about reproduction. The autonomy of the uterus and the autonomy of the person with the uterus have acted in concert for hundreds of thousands of years.

That autonomy is under threat. The racist practice of fertility control, or the "rational planning of future populations" to avoid "degeneration," has been part of our political climate for almost two hundred years.[8] The unreasonable, unscientific desire to control uteruses has led to some in the United States outlawing abortion for ectopic pregnancy, a condition in which embryo implantation occurs in the fallopian tube or outside the uterus and leads to death for parent and fetus.[9] Some people would literally rather have a fallopian tube explode and have a pregnant person die from internal bleeding than let them have a lifesaving abortion. Other state laws so fundamentally misunderstand early miscarriage that, technically, they require an official burial for menstrual blood.[10]

The political Right is arguing from a position of a naturalistic fallacy—that what happens in nature is good (perhaps they are using phrasing like "God's will," but it amounts to the same thing), and therefore any pregnancy, whether it is wanted or unwanted or life-threatening, should be forced to continue. Their position on uteruses is impinging on the bodily autonomy of those with uteruses. The way these politicians, fueled by money from special interest groups, steamroll onward with abortion omnibus bills, trigger laws (antiabortion laws that can take effect now that *Roe vs. Wade* is overturned), and more have killed people. They will continue to kill people. This is intentional.

When it comes to understanding the uterus, there are two core scientific principles: that it will vary in response to stressors and that the agency with which it acts only works in concert with the agency of the person housing that uterus. Our bodies are built for choice, and people without uteruses have been jealous of and have wanted to exert power over them for a long, long time. Many people are disgusted by menstruation because they have been taught to be disgusted because disgust leads to silence, and the more everyone stays quiet about the wonders, the spectacles, the generative power of this organ, the more those of us with uteruses cede of our agency.

Menstruation is something that half the world does for a week at a time, for months and years on end. Yet menstruation is stigmatized, largely erased from public life. In the pages that follow, I will show how knowledge was stolen and withheld from wise women in order to boost the status of professional men seeking to legitimize the new field of gynecology, how the cultural and historical underpinnings of the study of menstruation come from race science, and how a feminist orientation to menstruation science is needed to expose and address these histories. I hope to deepen your view on periods and make apparent their rightful place as one of the most wild, captivating biological processes in the human body.

MENSTRUAL MYSTERIES AND FERTILITY CONTROL

When I completed that project on periods and iron, I did not realize that it would be the first of many times I found an assumption was based more on bluster than scientific scrutiny. I did not expect menstruation to be so mysterious. Menstruation is not inscrutable or enigmatic because it is magical, though I will talk later in this book about how many cultures consider it sacred, and for good reason. It has to do with both the history of gynecology and with what people with money decide is important to fund; that is, it has to do with who controls the scientific study and medical care of the uterus.

For centuries, knowledge about menstruation, pregnancy, childbirth, contraception, and abortion was developed and passed among those who had uteruses, mostly women. And in many cultures, that is still the case. Yet a paternalistic perspective that grew dominant as gynecology took root is that knowledge of the uterus (and anything else worthy of scientific study) should be held among people with a particular kind of formal education and authority. As the historian Dr. Deidre Cooper Owens has shown, for well over a century gynecology was dominated by white men who sought to gain credibility and professionalization by operating on unanesthetized enslaved Black women, by working with and learning from these same women without ever crediting their knowledge, and by pushing out the midwives who had been doing this job in the first place—often Black and brown women.[11] There are brave examples of people who tried to subvert efforts to criminalize professional midwifery, yet over time the hoarding of knowledge among professional gynecologists was a successful venture. Discouraging and even forbidding passing on knowledge, creating community, and caring for each other between people with uteruses was formalized in many ways by early (and still by contemporary) medical societies and hospitals to control who could practice medicine.[12] There is a long line through the history of dominant science being practiced by titled men of the gentry, property-owning men, and now professional men that continues into the modern day. For some time now, in public and private life, uteruses have been under the control of people who tend not to have them.

Even the people who are first-line medical providers for children—children who could use support as they encounter menstruation for the first time—are inadequately, almost embarrassingly, unprepared. A recent survey assessed pediatricians' knowledge and clinical practice around menstruation, including whether these doctors discussed menstruation, menstrual practices, or menstrual health with their patients.[13] Fewer than half of respondents knew when in puberty menarche (that is, the first menstrual period) happens, how long menses lasts, and even how long it is safe to wear a tampon. Almost 15 percent of women pediatricians and 34 percent of men pediatricians surveyed provided an incorrect answer to the question "Can girls/women swim in the ocean with a tampon inserted?" (The correct answer is yes. Yes, they can.)

The mysteries around menstruation, then, are mysteries because the people who hold the positional, financial, or cultural power in science and medicine have for a long time asked a limited set of questions, and employed a limited set of methodologies, that are interesting only to them in their desire to maintain power and control over fertility. At first look menstruation does not fit into our understanding of fertility and reproduction because the framework with which we have thought about why menstruation happens, and how, is one that has long viewed the biology of people with uteruses as a set of peculiar abnormalities.

When I examined the literature about periods while I was in college and graduate school, I found the most popular hypothesis for the evolution of menstruation is, effectively, that it didn't evolve.[14] Instead, the reason humans menstruate has to do with a basic physiological phenomenon wherein once cells in the body hit their final, most specific form, they are counting down to their deaths. Endometrial cells do this to prepare for a possible pregnancy. So according to this hypothesis and its adherents, menstrual tissue is simply tissue that has hit its expiration date. We menstruate because we need to get rid of this tissue to start a new cycle and try again (so really, not so far from the "failed baby" story I got as a tween). At about the same time that the "periods are useless" hypothesis was proposed, there were two women putting forward their own different hypotheses suggesting menstruation is meaningful: one to rid the body of pathogens, the other to conserve energy.[15] Why had the women's hypotheses gained far less traction than this other one?

I suspect the reason it was so easy to accept the "useless" hypothesis and reject the ones arguing for menstruation's functionality has to do with the common refrain through the history of anthropology: female bodies and behavior are boring. Primate males (including men) are dynamic because they are violent and competitive.[16] Females? Well, we don't know much about what they do. They take care of the babies, I guess? It took decades for anthropologists to center the important research being done to recognize the value in female friendships, domestic labor, and care work.[17] This is what happens when you just keep building upon a scientific foundation that used to think women were a secondary version of men and that vaginas were inverted penises.[18]

In this book, we will look at that foundation. And most of the time, we'll blow it up and start over because it's pretty difficult to build good knowledge on top of bad. The normal, healthy menstrual cycles so often detailed in medical textbooks and health classes are framed as static, twenty-eight-day phenomena: the reality is they are malleable, responsive, dynamic. Menstrual cycles, and therefore ovarian and uterine function, must be variable and responsive in order to determine whether conditions are good enough (or not) for reproduction. The investment of nine months of gestation and several years of lactation, not to mention overlapping dependent offspring, is no joke. And if conditions do not support this level of investment? Better to wait. There are multiple points at which the body can decide conditions are not right: at ovulation, at gamete fusion, when the trophoblast interfaces with the endometrium, or when it begins to dig in and get closer to the parental blood supply. At each of these points, the process of reproduction can and does end when it needs to.

These are autonomous decisions based on signals the body receives and interprets from the environment, even if they are not always the

choices we wish to make (meaning, sometimes we get pregnant when
we do not want to, or sometimes a wished-for pregnancy does not happen). Physiology and behavior coevolve: our bodies might have ways of
understanding whether we have the resources to reproduce, but our
minds provide additional input, and strategy, regarding the timing and
tempo of major life events. In this way body and mind are not separable
but on the same team.

Essential to how we understand evolution is the key word "selection."
Natural selection and sexual selection are processes not just about adaptation but about what an individual does to shape their life within
that environment, from the tactics they employ to the relationships they
tend. Choice is fundamental to evolution: people from multiple cultures
choose to avoid conceptive sex for a period of time after the birth of a
baby to avoid children spaced too closely together and use contraception and abortifacients to avoid or end pregnancies.[19] Our bodies and
our minds take in different signals to together draw conclusions about
the desirability of reproduction at any point in time.

The variability and ways in which the ovaries and uterus respond to
environment form the core of this book because this variability contradicts the common medical lens of the single normative menstrual cycle.
This concept of normativity comes straight from eugenic science, and
it is a misplaced belief that those who occupy the averages are healthy,
and those on the margins are flawed. Rather than eugenic science oc-
cupying some small niche of anthropology, in my reading it's clear that
it's eugenics all the way down.

EUGENICS ALL THE WAY DOWN

I am an anthropologist, and I do love anthropology. But my field is not
without its problems, as its roots are in race science. Anthropology was
(is?) deeply connected to the activities of seventeenth- and eighteenth-
century European travelers who used their observations of the world to
fine-tune their theories about human nature.[20] Many of the most con-
sequential and lasting ideas about race were also about gender, and vice
versa; a particularly enduring idea has been that gender differences are

more pronounced among Europeans compared to Africans and that civilization and being of a superior race are what produce masculine men and feminine women. White femininity is therefore held up as a model toward which women of other races are expected to conform to survive within the civilizing forces of colonialism.

A furtherance of these ideas came from eugenics in the nineteenth century, the concept developed by Charles Darwin's cousin Francis Galton, who was interested in the cultivation of the English race and other "high races." Galton wished to use science to promote the selective breeding and careful improvement of people with European ancestry to maintain their dominance in the world.[21] Like Darwin on the *Beagle*, early natural history replicated and reinforced colonial hierarchies in that white people wrote expedition observations regarding the peoples and places where they traveled that centered their perspectives and discoveries (and sometimes even made these remarks without actual observations). Studying the other was a study in objectivity, so the thinking went, and of course Europeans were the ones who should extract, discover, and pioneer in already populated lands, among people who already had their own science and own knowledge. This clarity of vision came from being part of a civilized and civilizing society and a hierarchy that kept Europeans and settlers of European ancestry at the top.

Of course, the way that power operates, especially when it comes to white supremacy, is that white people are the ones who dictate cultural values, which means that white femininity is the right kind of femininity, that white bodies are more valuable than Black, brown, or Indigenous bodies, and that the white social structure is the pinnacle of human achievement. For centuries, these notions have dictated how we think about evolution, natural selection, sexual selection, and reproduction. The connections between anthropology and eugenics, and gynecology and eugenics, are important to expose because the underlying values that drove the research questions that led to contemporary understandings of the uterus obscured the reality of what this rather marvelous organ does.

When white femininity is centered as the best kind of femininity, a few ideas about gender get embedded into our science. Normative white femininity is often passive and in service to white masculinity, so we develop ideas about eggs as princesses in a tower and sperm as rescuing princes.[22] Normative white femininity is thin, so fatness harms reproduction.[23] Normative white femininity operates in strong contrast to white masculinity, so a butch, nonbinary, or transmasculine menstruating person is inconceivable and receives delayed or inadequate care.[24] As it turns out, eggs are the dominant ones (in fact, the whole reproductive tract gets in on sperm selection), eating too little is the greater harm to reproduction, and plenty of people of all genders continue to menstruate and even desire pregnancy in equal proportion to cisgender people.[25]

Notice that I equated "normative" with "best" here. Eugenicists had a direct role in linking these concepts, which affects not just gender (and race, class, and sexuality) but health. In the early nineteenth century, much of Western European medicine considered every patient's struggle as unique, the work of doctors more art than science.[26] This kept accountability low; if you do not survive a doctor's treatment, it cannot be medical error because we don't know how many people typically survive any affliction. Seeing medicine only as art meant there was no standardization of care, leaving room for doctors to peddle their own cures. But soon the individualistic, authoritarian model of medicine somewhat gave way to standardization: of calculating averages and probabilities in order to determine best practices.[27] This is not all bad.

Yet the shift from art to standardization is also when the field of medicine started to equate certain statistical calculations, like normality, with health. Normal eventually came to be associated with the *normal distribution*, in particular the data that cluster in the middle, or the average. This definition of normality impacts how we understand menstrual cycles and their variability. A lot of factors relevant to the menstrual cycle are, guess what, *not normally distributed*. There is no cluster in the middle, so any calculation of average is not an actual calculation of normal. So if in medicine we tend to calculate that those who fall within the normal are healthy, a healthy menstrual cycle is by definition difficult to achieve. The idea that there can be a normative menstrual cycle goes against the contextual menstrual cycle that we actually see out in the wild, and this equation of normal with health is part of the reason so many of us with uteruses develop ideas that we are somehow abnormal. Trying to repair this misapprehension of normality is a large part of what drives my lab's research.

The final point about eugenics that is key to this book is that eugenics has brought about a kind of medical paternalism, as a system that is upheld and reinforced by those who benefit from it. On the one hand, medicine cares deeply about preserving fertility, particularly of white women. On the other hand, medicine effectively gaslights people who express pelvic concerns, especially relating to pain or excessive bleeding, especially if those patients do not adhere to white normative femininity. This undervaluing of the lives of people dealing with illness, disability, and/or gender dysphoria lies at the heart of the medical betrayal experienced by many people with uteruses who seek care. Preservation of white fertility has been the goal for a long time: the earliest attacks on abortion in the United States were in the mid-nineteenth century, and physicians at the time made clear that their concern was in the use of abortion by white women. They argued that "middle-class Anglo-Saxon married women were those obtaining abortions, and that their use of abortion to curtail childbearing threatened the Anglo-Saxon race."[28]

Fertility control is a central tenet of eugenics. As the historians Dr. Susanne Klausen and Dr. Alison Bashford show, feminism and eugenics fell in and out of alliances as both favored the creation of contraceptive technologies. However, as Klausen and Bashford note, "Just as feminism was starting to promote voluntary motherhood, eugenics was arguing that women's reproductive capacity was too important to the race/nation to be left in women's charge."[29] Control over the uterus is evident in a lot of modern medical care and antiabortion activism today, just as it was then: Black patients are more likely than white patients to be encouraged to use long-acting forms of contraception (like intrauterine devices, or IUDs) even though scheduling IUD removal can be challenging and even met with resistance by one's provider.[30] Black patients, as well as disabled patients, have had to endure disingenuous allyship by antiabortion activists (including Supreme Court justices) as these bad actors chip away at reproductive rights with trait-selection laws (for example, banning abortion on the basis of race, sex, or ability status).[31]

This "Daddy Knows Best" model of reproductive control limits access to assisted reproductive technologies, contraception, and abortion to those with a uterus under specific conditions.

Our cultural climate affects my day-to-day, and it affects my science. Yet part of a feminist approach to the responsible conduct of research is reckoning with our history and our contemporary conditions and the responsibility we scientists often bear as people with authority. Many modern anthropologists, especially Black, Indigenous, and Latinx women scholars, have worked to forge a future for anthropology that problematizes colonial and race science rather than advances it.[32] These anthropologists do not ignore our history but use it to understand why expertise is assumed in those with power, why that leads to people thinking one person can be the expert on another's lived experience. They ask not only about quantitative evidence but about affective responses. These scholars are why I can write a book like this today that has any hope of disrupting the race science that has undergirded the study of the uterus almost from its inception. I want to get to the reality of how this organ works. Therefore, to do the work well, menstruation science requires a feminist approach.

This feminist approach must acknowledge histories and positional power and more and be careful with the language we use. Given the power I hold—I am white and cisgender (my gender aligns with what I was assigned at birth); I teach students, thereby controlling their academic success; I have tenure, thereby voting on employment decisions of more junior faculty; I'm writing this book; and much more—it's important that I am clear about my methods and my language because this specificity can help make menstruation science inclusive, can help notice inequity, and can point toward justice. This is where I am grateful to be both a feminist scientist and an anthropologist.

WHAT IT MEANS TO WRITE A FEMINIST SCIENCE BOOK ABOUT PERIODS

For the last fifteen years, I have led a feminist science lab that studies the environmental stressors that can affect the menstrual cycle. We have looked at how childhood, physical activity, diet, immune challenges, inflammation, and more affect ovarian hormones and how they in turn affect the uterus. The menstrual cycle, and menstruation in particular, is a fundamental biological phenomenon that happens to about half of us and is relevant to the continuation of our species.

I call the particular way we study the menstrual cycle in my lab feminist because of its practices as well as its outcomes. Feminist methodology is deceptively simple: uncover history, look to who holds power, and test the assumptions that tend to underlie it all. Feminist methodology in science requires one to be in conversation with social scientists, historians, and science and technology studies scholars to trace an idea backward and hear the stories of the people and ideas that were dominant at the time, as well as the views of those who spoke against that dominant thinking. Feminist methodology in biology, especially human biology, requires paying special attention to funder, researcher, and subject, to whose knowledge is valued, to what research aims get proposed, to what results end up in the peer-reviewed literature, and to what ideas and experiences persist even when supposedly rigorous science has ruled them out.

It also helps to be an anthropologist. Thanks to the feminist practitioners who came before me, I am an anthropologist who has been taught to pay attention to lived experience, and I try to withhold my own judgment or expert opinion on a topic until I have spent a lot of time listening. In our lab, we endeavor to perform what the anthropologist Dr. Anna Tsing calls the "art of noticing."[33] Tsing uses the metaphor of polyphonic music, multiple independent melodies played together, to help us understand that in order to do the work of noticing we have to be willing to step back from the idea that there is one dominant melody and "listen for the moments of harmony and dissonance they created together."[34] In her book *Against Purity: Living Ethically in Compromised Times*, the philosopher Dr. Alexis Shotwell steps into this conversation on noticing and adds that "to notice the world [is] a practice of responsibility."[35] Noticing allows us to see power and to follow multiple threads without allowing one to overshadow the others. Noticing allows us to see variability, and subversion, and resistance.

I've talked about feminist practices, but I also want to say something about feminist outcomes. A feminist outcome in science is about repairing harms and seeking justice. Does this mean I'm biased? In a way, heck yeah! Thanks to feminist and restorative forms of science, I know that when I am dealing with a lot of stress and my stomach hurts, this is a psychosomatic response, which does not render my symptoms imagined but acknowledges connections between the gut and the brain. I know that when I am in labor, certain procedures that may be dictated to me by my practitioner often beget more procedures and lead me almost inevitably to a cesarean section (notice I am not devaluing cesarean sections here but simply pointing out how certain systems produce certain outcomes). I don't know if any of the people who did the work that led to this knowledge consider themselves feminist scientists, but I know the result of this work has empowered me as a person who desires knowledge, respect, and autonomy.

Because of my anthropology training, because of feminist methodologies, my approach to menstrual science is very different from a clinician's or an epidemiologist's. For instance, I think we over rely on the idea that randomized controlled trials are some sort of charmed methodology that verifies the truth of a medical claim. I am not interested in recruiting homogeneous data sets in the interest of controlling variation because I wonder in doing so who our findings represent. My research questions are not oriented around a particular disease state. Instead, I'm curious about the whole wide range of what it means to be a person who bleeds. And I like to learn directly from people who have the experiences I'm curious about and see what they think is going on.

My curiosity and openness have led to conversations with friends getting real intimate real quick, social media messages involving screenshots of data from period tracker apps, long email screeds from both doctors and concerned parents, and in one instance, a stranger sending me pictures of their blood clots hanging off a tampon. Like many other feminist scientists, I'm pretty committed to the idea that the people with the lived experiences are the experts on said experiences . . . even if it sometimes means a surprise iPhone photo attachment.

As I wrote this book, I kept being confronted with my own untested assumptions, my own logical fallacies, and times when scientists so thoroughly failed menstruating people that I had to do a deep dive into the history and cultural context of a topic to understand why. I love science and the scientific method as the way that I make meaning and sense of the natural world. And yet to my mind, to truly love science means looking as much at its harms as its revelations. It means holding scientists and the broader scientific community accountable so that we can do our best work. I hope to take the mysteries out of menstruation without losing the magic.

HEY, YOU GOT YOUR SEX IN MY GENDER!

While I hope it is obvious that many people who are not women menstruate, and that many who are women do not, menstruation research often seems unaware of this fact. In addition to children who menstruate, and trans men and nonbinary people who menstruate, there are women and postmenopausal people who do not (and of course, plenty of people who occupy multiple of these categories at once, like my transmasculine teenager). Therefore, I have tried throughout this book to be very clear about what I mean when I use certain terms.

When I am discussing papers about girls or women, I am going to use terms like "girls" or "women"—these terms include anyone who is a girl or woman regardless of their assigned gender or sex. When I am expanding a paper's findings and suggesting it applies to all menstruating people, I am going to use terms like "menstruating people." It's important to be specific because some research is relevant to people with uteruses whether or not they menstruate now, and of course there are people whose uteruses are quiescent because they are postmenopausal but in their history have had hundreds of periods. So you will see the word "women" some, when it applies, along with people with uteruses, people who menstruate, former menstruating people, postmenopausal people, and more, depending on what I am talking about. Often I will simply describe the relevant body part, organ, mechanism, or function without feeling the need to explain the human being the part inhabits. Menstruation contains multitudes!

That's not to say menstruation is not without gender, given what I said above about how healthy reproduction is tied to normative white femininity. Gender and other social categories influence experiences and knowledge around menstruation, which is why it's worth taking a moment to clarify what sex and gender are as they relate to how we understand menstruation.

If you are Gen X or older, you may remember the old Reese's Peanut Butter Cup commercial in which a guy is strolling down the street, enjoying a giant bar of chocolate, and bumps into a woman enjoying a giant plastic tub of peanut butter (can we normalize carrying around and eating directly out of tubs of peanut butter please?). They simultaneously exclaim, "Hey, you got your chocolate in my peanut butter!," and "Hey, you got your peanut butter in my chocolate!" Sex and gender are kind of like that: a delicious, disentangle-able combination of traits, experiences, assumptions, and resistances. Sex is a designation that often has to do with a person's gonads, genitals, and/or sex chromosomes, whereas gender refers more to shared cultural experiences or identity.[36] Yet both are neither solely biological nor solely cultural, and it would be scientifically inaccurate to try to make categories for sex or gender binary.

The social psychologist Dr. Suzanne Kessler shows how arbitrary and nonbiological sex designations can be in her interviews in the 1980s with surgeons who performed genital surgeries on intersex infants (these surgeries on nonconsenting infants are now, thankfully, rare). Urologists who "like to make boys" see a child who needs a penis; at the same time, what can dictate gender assignment surgery for an infant with a small penis is "whether it is 'good enough' to remain one."[37] Gender assignment of infants is a combination of bias and art, not science, just like those nineteenth-century doctors who did not want statistics meddling in their profession. One's gonads or outward-appearing genitals do not "match" one's chromosomes nearly as much as we assume because of intersex erasure; some studies suggest as many as 2 percent of infants are born with intersex traits.[38]

An awareness of the biological shortcomings of prenatal or infant sex assignment is one of the reasons many scientists are moving toward terms like "gender/sex" to describe a phenomenon of expression, biology, culture, and lived experience that cannot be disentangled.[39] While

I will use "gender/sex" when it applies, I will also use the term "gender" when I am referring specifically to cultural forces acting on the body. As described above, gender is heavily influenced by race. That is, in the United States and many other countries the idealized forms of gender roles and behaviors are often drawn from what we as a culture want and expect white women to do and are in part intended to distinguish white women from women in other racial categories.[40]

Ultimately, while *gender* may have some underlying biology because it is partly correlated with chromosomes, gonads, and genitals, *gender inequality* is a series of social practices unrelated to biology that has biological consequences. This means a menstruating person's gender presentation, gender beliefs, and experiences of their gender will influence the experience of something as gendered as menstruation and have consequences for how they interact with the world. And when it comes to bathrooms or other ways that people experience menstruation in public, those consequences can be ones of real physical danger.

In addition to issues of gender/sex, there are a few sneaky terms I need to clarify. I'm talking about the anthropologist's propensity to talk about Western cultures as though they are normative and to hide lots of other concepts inside the term.[41] When I use the term "Western" in this book, I am referring to history, culture, and/or science as they derive from Western Europe. The history of the West (as well as how we sometimes imagine that history, true or not) plays a strong role in much of Western science. However, something that can also happen in anthropology is that when we say Western, we are also sneaking a little bit of race science into the mix, meaning that when we say Western, we are also conflating it with whiteness. This happens not just in concepts of Westernness but in the makeup and practice of Western science: most Western science is done by white people, and most Western science intended to establish universals or normative values is done on white people. As with gender/sex, specificity is key. Yet unlike gender/sex, we need to do the work to separate out these terms so that the impacts of race are made visible.

When I say white I am referring to ways whiteness is often positioned at the top of the racial hierarchy, which is of course enmeshed within a lot of Western history, culture, and science. But it needs to get pulled out and specifically named more often than not, not only to notice how racism is acting on a given concept but to keep from erasing all the Black, brown, and Indigenous people who contribute to Western culture, who transform it, who resist it. Finally, when we are talking about "Western," the other major conflation that happens is that we often also mean an industrial or postindustrial environment—living in cities or suburbs, lots of white-collar jobs, certain concepts and expectations around formal education, diet, and physical activity. But a population's subsistence activities are more heterogeneous, as well as the impacts of those activities on the body depending on gender, race, and class. So when relevant, yes, I'll talk about postindustrial environments and industrial environments, as well as farming, foraging, and the like. But none of these activities or broader cultural practices are Western versus non-Western.

Finally, with great appreciation I want to note the geographer Dr. Max Liboiron's concept of dominant science, which they introduce in their book *Pollution Is Colonialism*.[42] Liboiron points out that dominant science is defined as science that "becomes dominant to the point that other ways of knowing, doing, and being are deemed illegitimate or are erased." While much of Western science dominates, to conflate the two misses there are some Western sciences that are not dominant, some that have been overpowered by other Western sciences, and some that Western scientists question regarding what has been and is dominant. The term "dominant science" is useful in the ways it allows us to notice when one melody overwhelms and invites us to ask what ways of knowing we might uncover if we listen closely.

I JUST WANT TO DO MY SCIENCE

If it sounds like I'm getting political in this introduction, it's because I am. Science is performed by humans who live in society.[43] Medicine is administered by humans who have their own histories. I don't want to be talking about this stuff; I want to be nerding out on mechanisms of menstrual repair. I don't want to be Googling which states have trigger laws; I want to look at estradiol curves. I promise that this book is full of science. It's also full of the history and systems at play that complicate the ability of menstruating people to get good information about their own bodies, ask their own questions, and be the ones setting the research agenda. This is tricky terrain because, in addition to the ways in which information has been withheld in the medical realm, in the political realm we see people who could use accurate information about uteruses to inform policy, and they choose not to, and that's kind of the point.

In the first few chapters, I will be taking on the why and how of menstruation. I will spend time on the concept of the menstrual taboo, how variably different cultures theorize about menstruation, how these things together led us to our twentieth-century view of periods as useless . . . and what a twenty-first-century reckoning does to modernize our understanding. I will look at the history of normality in science and medicine and show how the desire to conceive of normal, healthy people narrowly has influenced our understanding of the menstrual cycle. I'll cover ovarian hormones, feedback loops, and the endometrium, but I promise I'm a good guide. I'll share the latest research out of our lab revealing that the "normal menstrual cycle" does not represent the majority of individuals and what we are doing to show medicine how to better characterize variation.

Next are three chapters that look at the sorts of environmental factors that explain the majority of variation in menstrual cycles. Chapter 3 starts with the story of the limited energy theory, the hypothesis put forward toward the end of the nineteenth century that too much of any activity will detract from a white woman's ability to reproduce (especially mental activity, like the vulgar desires to go to college or vote). I also take on energetics: how much we eat and how much we move around, as well as how some of our misconceptions around food and body size lead to major misunderstandings around menstrual cycles and some of their pathologies, like polycystic ovary syndrome.

Chapter 4 covers immune challenges. I look at how inflammation can get in the way of signaling in the uterus and what that means for menstrual cycles. I also spend a lot of time on two phenomena close to my heart. The first is the concept of *menstrual hygiene management*, MHM for short, a public health intervention often intended to address period poverty issues among menstruating people in the Global South. One of the rationales behind MHM is that many methods menstruating people are forced to use to make their menstrual blood invisible to others increase infection risk. But is the menstrual blood (and associated rags, straw, or newspaper) what is dirty here? Next, I'll discuss a project that recently took over our lives. Our project highlighting the heavy and breakthrough bleeding experienced by many after receiving a SARS-CoV-2 vaccine has ended up taking on issues of vaccine hesitancy, medical mistrust . . . and how vaccine trials were not designed to look at menstruation.

Chapter 5 covers the final major category of environmental factors that can affect menstrual cycles. Psychosocial stressors have an interesting place in the history of medicine: we have bouts of hysteria, nerves; we are tossed to and fro by our premenstrual phases. At the same time, I have met a lot of resistance over the years in my field when anyone tries to contend that psychosocial stressors may influence menstrual cycles. It's as though menstruating people (especially those who are women or feminine of center) are intensely susceptible to stress, yet it stays in our heads and is unlikely to become embodied in any meaningful way. I do not buy it.

Finally, in chapter 6 I take on the future of periods. Should we just suppress our periods indefinitely? Work toward a more accommodating culture for menstruating people? Science and technology studies and anthropology have had a little to say about this over the years—and I imagine you have your own thoughts about this too. How much of the hassle of periods is because they are a hassle, and how much is because the structures that underlie family, home, work, and being out in the world make them a hassle? What if it's okay to hate periods, okay to like them, okay to be largely ambivalent about them? A feminist future is going to have to make room for all of these perspectives, and an intersectional feminist future needs to grapple with not only what all of this has to do with gender but what it has to do with race, colonialism, class, heterosexism, and more. What does it look like for menstruating and nonmenstruating people to be in community with one another, to recognize the interdependence of all people and begin to take some responsibility for each other?

The boundaries of what is allowable in our period future—contraception, menstrual suppression, abortion, prenatal and postnatal care, medical treatments, and more—are constrained by what is known and who does the talking. Since the same constituencies telling us what we can do with our uteruses are the same as those who have dictated what we know about them, we have a bit of a problem. But now you have this book. With what we know, what different futures might we imagine, and how might that fuel our efforts at societal transformation?

In *Hope in the Dark: Untold Histories, Wild Possibilities*, Rebecca Solnit writes that "to hope is to give yourself to the future, and that commitment to the future makes the present inhabitable." Hope has to be active to help us imagine and work for a better future. So I look ahead to the rest of the book with hope, in that activist sense of the term. My scholarly training happened not only in the classroom, the lab, and the field but at civil disobedience trainings and on picket lines. Some of my best teachers in graduate school were my fellow labor organizers because I learned that you have to get up quick after failure and get ready for the next battle. Being a part of the fight was what made the present inhabitable for me. And to me, this book is part of the fight for choice, for health care, for reproductive justice.

I hope that you will read this book with the eagerness with which I wrote it, that you will gobble up my descriptions of what uteruses can do, that you will revel in this organ's agency and adaptability, and that you will at a minimum develop a grudging respect for menstruation and the mechanisms of menstrual repair. I hope you will see how inextricable the connections between power, gender, and race are when it comes to exposing menstrual misconceptions. The purpose of this book is not to fill a knowledge gap so much as to show you what is possible; to give hope, to act on hope, that what our bodies produce and excrete and transform matters and that to care for our material body is a radical act. I want these words to provide fertile ground for new ideas of what a world could look like where we demand and fight for our bodily autonomy and imagine radical futures.

CHAPTER 1

There Is a Reason for All of This

SWAPPING OUT a laden pad on my heaviest flow day, my back or vulva or abdomen aching, I often found myself wondering about the point of it all. Why do we menstruate? Why do we do it so very often, over so many years? And why is it such a hassle? Beyond lamenting the physical discomfort—I used to experience such debilitating pain during the first day or two of my menses that I'd have to cancel labwork and classes—I was irritated by the inconvenience: the need to predict when my period might come, to anticipate the supplies I might need, and to figure out whether I needed to leave a meeting early or risk staining my pants.

Not everyone feels this way about their periods. Not everyone, of course, experiences pain or discomfort with them. What I mean, though, is that not everyone has the perspective that menstruation is a hassle or inconvenience. Menstruation, in their view, is not something to hide or endure while carrying on with one's normal routine. It can be, instead, something to pause and experience, to acknowledge rather than survive. Dr. Whitney Robinson, an epidemiologist who studies racial, ethnic, and gender disparities in reproductive health, points out that there are plenty of countertraditions to the dominant narrative, from the "strong, earthy woman" to Black magazines writing about conditions that differentially affect their communities rather than stay silent.

The perspective we take on menstruation, and therefore how we directly experience it, is strongly tied to our culture. How we have to handle menstruation in public, and how it is publicly perceived, is going to affect our own relationships to it. The way Western culture values maleness and whiteness shows up in two of the most important and enduring public experiences we have: school and work. As students, our school norms often promote rather than dismantle inequality and oppression, and what's more, menstruation can invite additional scrutiny (as well as judgment and disgust) as one has to navigate gym class, bathroom passes, and the physical distress that can accompany bleeding.[1] Some school curricula also lack an empathic frame that makes it possible for children to expand their understanding of who is worthy of dignity beyond their own strict lived experience.[2] Workplaces often continue this trend: male-dominated workplaces in particular tend to have higher rates of bullying and harassment, encourage competition and workaholism, and prize strength over weakness.[3] Behavioral norms commonly attributed to whiteness, meanwhile, include a constant sense of urgency, perfectionism, and the protection of the emotional comfort of the powerful.[4] These characteristics put pressure on everyone—but especially those historically and currently marginalized by school and work structures—to hide differences often perceived as weaknesses in order to survive. These public experiences in turn have a strong influence on the way we understand our own experiences.

It was only when I began to question why I saw menstruation as such a hassle that I started to notice that I was not alone in holding this perspective. It was also held by the science that I had been reading about reproduction for my entire life. English-language Western science is conducted primarily by white people and by men. Anthropology in particular is rife with hero myths—journeys our species goes on over the course of human evolution to emerge victorious as white male *Homo sapiens*.[5] We see these hero myths in anthropologist Johann Friedrich Blumenbach's five races in the eighteenth century, where "Caucasians" were the pinnacle of enlightenment but all other races degenerate; medical doctor Franz Joseph Gall's phrenology that followed and just so happened to establish differences between races and sexes that reaffirmed existing hierarchies of worth; anthropologist Raymond Dart's twentieth-century hypothesis that our early ancestors were warriors at the top of the food chain, eating meat and fighting each other with broken bones (for this last, it turns out that what he thought were weapons were predator assemblages indicating that we, in fact, were the prey).[6]

Even when science is conducted by people who don't hold these identities, the white male perspective is upheld and valorized as the objective one, what the philosopher Thomas Nagel popularized as the "view from nowhere."[7] For a time in the 1990s, nonwhite anthropologists had to label their work "native anthropology" to differentiate it from the supposedly more rigorous work of white anthropologists who study brown people; brown people could surely not have the appropriate distance to be objective about their own culture, you see.[8] Yet the more I have learned about menstruation, the clearer it has become that Western science's view on menstruation is very much a view from somewhere, a dominant paradigm increasingly imposed throughout the world that is ignorant of women, anyone not cisgender, people of color, and/or people with disabilities.

With this foundation in mind, as I explore the question of why we menstruate I begin by considering some preliminary answers from cultures and traditions around the globe before discussing what Western scientists think about it. Showing how menstrual taboos operate across several cultures will, I hope, make it easier to see how research in Western science has been colored by our own cultural context and belief systems. Some of our earliest hypotheses, tested using what we tend to assume is an objective scientific method, were clearly influenced by Western concepts of female pollution. Engaging with these multiple ways of knowing affords us the opportunity to step back from the dominant frame with which most of us have learned about our bodies and more clearly see the often-unreflective influence of this frame on what is disseminated all over the world under the name of objective science.

After considering how our varying cultural origin stories motivate our answers to why we menstruate, I will then disentangle the question of why menstruation happens from an evolutionary perspective from the question of what purpose it serves in processes of reproduction. To do this, I'll focus on the concept of *spontaneous cycles*, a biological phenomenon we share with a number of mammals that will help contextualize our menstrual cycles, and thus our periods. Finally, I will discuss new scientific evidence around an often-deadly disease of pregnancy that suggests that, rather than polluting the body, menstruation and its adjacent processes play an important role in saving lives.

MENSTRUAL TABOOS

Many older ethnographies and historical works describe menstrual taboos as phenomena, universal across cultures, aimed at protecting the community from the evils of menstruation. The purpose of menstruation is to shame, to serve as punishment for women's wickedness, or to purge the impure from the body, and menstrual taboos function to contain the pollution. These ideas persist in the Western popular imagination because they align with our own interpretation of menstruation. There are, of course, fewer places today where the norm is to seclude menstruating people or practice particular rituals to contain menstruation's polluting power. But the menstrual taboo has not disappeared; it has merely gone underground. In the United States, for instance, negative attitudes about menstruation appear in how people are taught to conceal when they are menstruating, to avoid talking about menstruation, and to reduce physical activity.

One book that disrupts the notion that menstrual taboos universally connoted pollution is *Blood Magic: The Anthropology of Menstruation*, edited by Drs. Thomas Buckley and Alma Gottlieb.[9] In their introduction, Buckley and Gottlieb make a strong case that many Western anthropologists, so convinced of their objective "view from nowhere," in fact projected their own belief systems onto the cultures of others. Menstrual taboos, they argue, vary widely in their meaning and often connote the power of menstruation rather than its evils.

The word "taboo," used in the English language to mean "banned on grounds of morality or taste," or even "banned as constituting a risk," bungles the original meaning.[10] "Taboo" comes from the Indigenous Polynesian term "tabu" or "tapu." For example, Dr. Mānuka Hēnare of the University of Auckland explains that for the Māori, tapu "expresses an understanding that once a thing is, it has within itself a real potency, mana. . . . Coupled with the potential for power is the idea of awe and sacredness, which commands respect and separateness. It is in this sense that tapu can mean restrictions and prohibitions."[11] Western colonists and anthropologists took on the term with only the narrowest focus on the idea of prohibition, entirely ignoring its connection to the sacredness of the material and nonmaterial world. With this context we can begin to see the rich and varied meanings that different cultures have applied to menstruation via varying forms of taboos.

One of the major restrictions often placed on menstruating people is that of seclusion from others. Given the long-standing Western assumptions about the nature of menstrual taboos, the most widely reported and remarked-upon forms of non-Western menstrual seclusion tend to be the most misogynistic and harmful. For this reason the most widely known menstrual seclusion practice among those of us consuming Western media is Chhaupadi in Nepal. In parts of the country that still practice Chhaupadi, when people menstruate, they and their young children are forced to stay in crude structures for the duration of menses.[12] This keeps menstruating people, who are often young parents, away from the rest of their families and their community for days at a time. A recent analysis of Chhaupadi among adolescent girls in the Achham district of Nepal, where some believe this practice originated, found that few of these sheds had toilet facilities, ventilation or windows, electricity, or bedding.[13] Nine of the seventy-seven girls in the sample who practiced Chhaupadi had been bitten by snakes while confined to their sheds. While Nepal has recently outlawed this practice, it persists due to a lack of enforcement, and menstruating people and their babies continue to die from exposure, snakebites, and smoke inhalation.[14]

In many cases, however, menstrual seclusion is more about protecting the power of menstruation and ensuring that it is not wasted on one's daily routines.[15] Among the Hupa of Northern California, for example, menstruating people gather in women's houses, called *min'ch*, which are also spaces used during the postpartum period and after a miscarriage.[16] Min'ch translates to "a small, familiar, or dear house," and their multipurpose use demonstrates that these are places not of isolation but of community. Dr. Cutcha Risling Baldy, a professor of Native

American studies and herself a member of the Hoopa Valley Tribe (the tribal government where many Hupa are enrolled), writes that "menstruating women were part of the luck that men of the tribe utilized when they were hunting or needed protection," which came from participation in the min'ch.

The Americas are not the only location where Indigenous menstruation practices are about protecting power. Among the Beng of Cote d'Ivoire, several menstrual taboos exist: a married or previously married menstruating person may not go into the forest (except to defecate), they may not touch a corpse, they may not touch the fire of a nonmenstruating woman, and they may not prepare food for men.[17] However, menstruating Beng are otherwise not restricted in their engagement with family, friends, and the major rituals of their community, nor restricted from sex or sleeping with their spouses. These taboos are intended to protect the fertile, menstruating person and provide separation between the sacredness of female fertility, which is represented by menstruation, and Earth fertility, which is considered male and represented by the forests and fields. Beng menstrual seclusion offers a chance to slow down, prepare tasty food, and avoid some of the labor that typically fills their days.[18] According to anthropologist Dr. Alma Gottlieb, Beng menstruators are known to make the best food, including a slow-cooked palm nut sauce, which they share with their female friends, sister wives, and kin. (Younger Beng men, less stodgy in their commitment to fortifying their masculinity, are quite willing to risk exposure to the sacred powers of the female for a mouthful of this sauce.)[19]

Another way to understand the sacredness of menstruation, and how different peoples have conceived of its purpose, is to look at coming-of-age and menarche (or first menstruation) ceremonies. Among the Asante of Ghana, menarche ceremonies celebrate the new menstruating person, where she "sits in public view beneath an umbrella (a symbol generally reserved for kings and other dignitaries), receiving gifts and congratulations and observing singing and dancing performed in her honor." Menstrual blood, on its own considered polluting in this culture, confers positive powers when used by priests to create magical objects out of brooms smeared with menstrual blood.

The menarche ceremony in Fiji involves newly menstruating adolescents staying indoors, at home, for four days. During this time the women in the adolescent's family explain to her what it means to be an adult woman: they tell her the mythology and history of their family, they tell her about menstruation and how it is not to be feared, and they tell her of the connection between menstruation and pregnancy, warning her to avoid sexual activity with boys until marriage.[20] In a recent study of this ceremony, participants represented it as a positive experience. A similar ceremony—in which the adolescent stays home for several days while the women in the household teach her much of the knowledge and skills she will need in adulthood—has been described for women in southern India. However, the author who studied this latter practice, a male social scientist from this same region, interprets these moments in an entirely different manner. As he writes, "That a woman's essential role in a patriarchal society is to bear children is reaffirmed on the occasion of first menstruation. . . . During this period the girl is fed with nutritious food, with the implicit intention of making the girl physically fit for biological reproduction and sexual life."[21]

In North America, Indigenous scholars and revitalizers of coming-of-age ceremonies demonstrate a range of possibilities for what menarche ceremonies can represent. The practices around menarche, some of which also involve periods of seclusion and education, are intended not to protect others from the new menstruator's pollution but to recognize the significant power she wields in this moment. Her performance of these practices is intended to help her use her power in a way that positively influences her own life course and the welfare of her community.[22] Many Indigenous peoples have had to resist Western, colonizing notions of femininity and in some cases imperial feminist attempts to delegitimize menarche ceremonies and other practices that white feminists see as regressive.[23] They have had to endure Indian boarding schools and community schools hosted by missionary nuns who disrupt these practices, forced assimilation into a culture that views menstruation as dirty, and physical harm and even death for preserving their own traditions.

Dr. Kim Anderson, a Metis scholar of Indigenous history, tells the story of Marie, an Ojibway participant historian who contributed to her book *Life Stages and Native Women: Memory, Teachings, and Story Medicine*. When she reached puberty, Marie began a ten-day seclusion, during which she fasted and experienced some restrictions on what she could touch and on her activities. However, her older brother came to visit after nine days because he heard she had stopped going to school on account of an illness. As Anderson recounts, "Marie lamented that she did not finish her seclusion, as she had just begun to dream and have visions of ancestral visitors." The transition to puberty, and later to menarche, signaled a transition not just to adulthood but to "the world of the spiritual." Being denied these practices, then, could be seen to have repercussions not only for her life but for the health of her community. In another instance, recorded by Dr. Thomas Buckley, a Yurok research participant described how she learned that menstruation is "bad and shameful" while living in foster homes. Upon returning to the Yurok, however, the adult women in her life restored her sense of its positive dimensions.[24]

Dr. Baldy, the aforementioned Hupa scholar and Indigenous ceremony revitalizer, describes her own similar experiences of menarche and menstruation. Being assimilated in the dominant US culture, she found it natural to see menstruation as something to hide and feel ashamed about. As she points out, "I felt no pride in menstruating. . . . There was no celebration of menstruation on TV; there were no moments where young girls talked fondly about their changing roles in life." So it is no surprise that when her mother offered to perform the flower dance upon her own first menstruation, she declined. Years later, at a vulnerable and difficult time in her life, her mother reminded her of this moment by saying, "Oh Cutcha, we should have danced for you." Dr. Baldy came to a realization:

> I was weak and tired, but that did not mean I was alone. There were songs echoing in my head, songs that I had heard growing up and songs that I had never heard before, and I thought about what it means when all of those voices sing together just for you. I sang to myself

that night, and that's what made me stand up, go to the bedroom, and pack a bag. I sang to myself as I walked out the door. I sang to myself as I got in the car. But I wasn't singing alone. I could hear the song in my head, and it echoed with voices from the many times I had heard my grandmother, my mother, and my aunties sing these songs to me as I was growing up. . . . I knew there was something important about our women's ceremonies. I knew that I would write about it and document the words of the women who had joined together to tell our girls that they are important, I would be reminded of this moment the first time I sang in a Flower Dance, the shakiness of my own voice calling out into the night. There I was, in my own way telling the young girl sitting just below me, "I am here. You will never be alone. We are dancing for you."[25]

These are just a handful of examples demonstrating the varied and diverse ways menstruation is represented across the world. For some cultures, the purpose of menstrual ceremonies is to contain and hasten the removal of the polluting influence of menstruation. For others, the purpose is to contain incredible creative powers in order to channel them in a way that best serves both the menstruating person and her community. Some societies appear to have conceptualized menstruation as a time for seclusion and retreat either because it offered an inbuilt opportunity for rest and for gaining spiritual insight or because in their view it was actually for said rest. That is, perhaps in some cultures, part of the reason for menstruation is its metaphysical function, and not only for the individual but also for the society. When we peel away Western and colonizing influences from the many stories told of menstruation, we might learn different ways of knowing, countertraditions, and new meanings.

MEDIEVAL AND EARLY MODERN VIEWS OF MENSTRUATION

Despite the examples we have just considered, much of the anthropological and biological literature suggests that menstruation is polluting, dirty in both a ritualistic and physical way. This connotation is strongest when the religion entwined with these rites was Judeo-Christian or Muslim or when the interpretation of the rites occurred through a Judeo-Christian lens, either because the interlocutor hailed from a country where that was the dominant worldview (as with many early anthropologists) or because the group had experienced forced conversion through encounters with missionaries.

The belief that menses is polluting may not, however, originate from source texts from those religions. The Old Testament offers guidance on appropriate cleaning practices for people who are menstruating as well as those who have ejaculated—the equality in treatment reflects the belief at the time that menstruation, like semen, was a material of conception. But that does not give one the sense that people who menstruate or menstruation itself is especially polluting. There is also a story in the New Testament about Jesus healing a woman many interpret to have been menorrhagic—meaning she had heavy menstrual bleeding. Rather than being kept apart, she was in a crowd of people who were all there to see him. She touched his clothes and was instantly cured. And while many often attribute negative attitudes toward menstruation in Judeo-Christian cultures to the fact that menstruation is "Eve's curse," Eve was not in fact cursed with a period. Eve's curse relates to pain with childbirth.

If the Bible is not the primary source material for Western conceptions of menstrual pollution, from where did they come? Historians of science identify the medieval period as the time during which the dominant attitude about menstruation may have shifted from one of simple restriction to a more hostile attitude toward menstruation's evil, polluting influences. Consider, for instance, the views put forward in *De secretis mulierum*, or *Women's Secrets*, the most widely circulated and translated medical textbook on women during this time.[26] Written by a monk who purported to be the more famous Albertus Magnus (but almost certainly was not), this textbook "presents the human being's emergence from blood as a secret of women's blood and bodies, about which men exchange medical knowledge."[27] In other words, the purpose of the book is to reveal the secrets of female bodies to men, though there is evidence that the author never treated women medically nor performed any autopsies on them.

Much of this text reads like an early conspiracy theory message board: the text is written in letter form as though to other men, inviting contributions from additional men, and suggests that only the most elite, in-the-know men are the intended recipients. Among the ideas that the text puts forward are that menstruating women give off harmful fumes that will "poison the eyes of children lying in their cradles by a glance" and, more broadly, that a "woman is not human, but a monster." This was not, however, the only view about menstruation held by the medical community during this era. Some believed it was simply a component of reproduction and material to aid in conception, others that it was undigested food because women's natural coldness compromised digestion. Older Galenic theories about the four humors remained popular. According to these views, menstruation is naturally occurring phlebotomy to get rid of plethora, or excess humor.[28]

Yet negative perspectives on menstruating people and menstruation began to crowd out more neutral perspectives. To understand better how this occurred, I spoke with Alanna Nissen, design manager at the Walker Art Gallery in Minneapolis and the author of an amazing master's thesis on witchcraft, the medieval period, and menstruation. As an art historian, Nissen was interested in concepts of the "monstrous feminine." Tracing the idea backward through time, she found that society's understanding of women dramatically changed during the early medieval period. One possible reason for this was the transformation from feudalism and the household production of goods to more of a capitalist, wage-based economy in which domestic, non-wage-earning labor, effectively the labor of women, no longer counted as "work." Another was the concentration of power within the Catholic Church. Taken together, Nissen points out that "this is also where you start to really see the idea of motherhood as a special vocation. And that redefines women as having a reproductive value rather than an economic value."

Nissen told me that brewing was once a female vocation but that women were eventually pushed out by men. Surviving in this profession became particularly dangerous because of the creation of the witchcraft stereotype, which intentionally incorporated aspects of the brewing profession—from the attire brewers wear to the stirring of cauldrons—as a way to exclude women. Witch trials were a means to control women, keep them out of wage work, and redefine them around their inherently dangerous, magical bodies.[29] It was during this period of "gender crisis" that theologians expanded Eve's curse to include not just childbirth but menstruation and that the consequence for transgressing prohibitions around menstruation came to include the penalty of death.[30] Menstrual blood, it was believed, was produced because women were unable to remove impurities using sweat, as was assumed for men. So menstrual blood came to be thought of as excrement, as filth, as polluting and foul. Only menstrual blood had the potential to "bewitch, deform, and kill."

As Western science continued to develop concepts around empiricism and the scientific method, I wish I could say that these ideas were quickly discarded. Unfortunately, for much of the twentieth century the prevailing research on the purpose of menstrual blood took for granted, and reinforced, these polluting and harmful beliefs. Strangely enough, the modern study of menstruation and menstrual blood starts with a satisfied patient, a menstruating lab attendant, and some wilted flowers.

A BRIEF HISTORY OF WESTERN MENSTRUATION SCIENCE

As the story goes, Dr. Béla Schick was a very popular doctor in the 1920s. Because he was so well loved, many patients would send him flowers, and he would often ask one of his lab attendants to put them in water for him. One day, however, a female lab attendant declined his request to do so. After being pressed about the matter, she admitted that she was menstruating and that when she handled flowers while menstruating, they wilted. In other versions of the story, she only admitted she was menstruating after Dr. Schick finds that the flowers have wilted.

To be honest, I might tell my boss something similar if he was asking me to put away a bouquet for him too: perhaps this lab attendant had better things to do. But Dr. Schick took her statement at face value and decided to run an experiment. He gently placed two bouquets of flowers in vases, one of which had been thoroughly handled by a menstruating woman. Should the handled flowers wilt faster, he argued, it would

The effect of the menotoxin on "cinerea." The
authors wrote, "In one case the flower was
placed in a solution of normal blood serum in
distilled water, while in the other the same
concentration of menstrual blood serum in
distilled water was employed; toxic effects
were noticeable within a few hours."

provide clear evidence that "some poison or toxin is present in the skin
secretion of the menstruating subjects which hastened the death of the
flowers."[31]

Over a series of short papers and correspondence in medical journals, Dr. Schick and others took up the cause of what they called the
menotoxin. Some injected menstrual blood into rodents to see if they
died, while others grew plants in venous blood from menstruating
women to see if it killed them.[32] As with the original experiment, the
menotoxin was not confined to menstrual blood per se but rather secreted in sweat, blood, breast milk, and menstrual blood during the
menstrual period. What's more, any menstruating person was now in
danger of being pathologized due to menstrual toxicity. One case study
reported that a mother gave her child asthma because she was meno-
toxic during pregnancy.[33] All manner of female ailments—and even
conditions experienced by those in proximity to menstruating
women—could be explained by the menotoxin. In the height of the
study of the menotoxin, Ashley Montagu—famed anthropologist,

outspoken critic of the concept of biological race, and popularizer of human evolution—dipped a toe into the menotoxin cesspool. After reviewing, and largely approving of, the science of menotoxicity, Montagu wrote the following:

> Twentieth century science appears at last to have discovered that menstruous women excrete substances which are capable of exerting a harmful effect upon living tissues of certain kinds. Primitive man has believed that women are so capable for countless centuries. Science, as the result of experimental investigation, attributes the capacity to the operation of certain chemical and physiological factors—primitive man to the operation of supernatural or magical ones.[34]

The belief in the scientific objectivity of the menotoxin persisted for several more decades. When I was an undergraduate, I spent several long afternoons in the basement of one of the science libraries, poring over correspondence between scientists in old issues of the *Lancet*. These letters, and additional articles on menotoxins, continued into the seventies. In one letter, an author relays her encounter with Dr. Schick just a few decades earlier, writing, "Dr. Schick and I discussed the possibility that the adult female diabetic out of control, the depressed adult female psychotic, and the adult female in the premenstrual phase secreted some common substance in their sweat."[35] It seems that, at least as late as the 1970s, we hadn't moved so far from the idea that menstrual blood can "bewitch, deform, and kill."

Though the concept of menotoxins eventually fell out of favor, the perspective that menstruation exists to purge something from the female body persisted in the scientific literature for several more decades. Two new, opposing hypotheses were published in the early 1990s. One postulated that menstruation is a mechanism to expel abnormal embryos, while the other held that menstruation expels sperm-borne pathogens. Few scientists supported the first hypothesis because abnormal embryos happen all the time to all sorts of species, and menstruation isn't necessary to get rid of them.

The second hypothesis represents the first time that men have been blamed for producing the "unclean substance" associated with menstruation, as best as I can tell. Margie Profet—a MacArthur award–winning physical scientist who produced several provocative hypotheses—proposed that menstruation is part of the typically female reproductive system's immune defenses. "Sperm are vectors of disease," Profet writes, an opening salvo that associates uncleanliness with testes, rather than uteruses.[36] In support of this position, she claimed that menstruation is universal among mammals but covert among some species, that menstruation is more copious among species with lower chances of pregnancy per cycle, and that menstruation is more copious among larger-bodied species or species with multiple mating partners.

Much of this evidence was eventually refuted in a 1996 paper by the anthropologist Dr. Beverly Strassmann. As Strassman observed, it's unclear what covert menstruation is or how it would be helpful in clearing infection from the body. If covert menstruation is simply the small amount of menses that is resorbed among what we typically consider to be nonmenstruating species, then it all stays in the body. It's also not clear that menstruation would be an especially useful way of ridding the body of infection, as it occurs often weeks after exposure to dirty sperm, and blood is great for promoting bacterial growth.[37] In fact, twentieth-century critics of the concept of the menotoxin pointed out that the deaths of rats from injections of menstrual blood were almost certainly from uncontrolled bacterial infections, rather than a mysterious menotoxin.

The second idea—that menstruation is more copious among bigger animals or more promiscuous ones—makes some sense. The bigger the body, the bigger the uterus; it would follow that a bigger uterus would make more endometrial tissue and thus be more likely to need to menstruate (rather than just resorb it). But even this isn't a perfect correlation. While this relationship more or less holds true among primates—the bigger animals, the apes, are the ones in which we see at least some menses—it does not extend to other mammals. Visible menstruation is present among several smaller mammals, such as bats, shrews, and the spiny mouse, and absent in plenty of big ones, including in one of our close relatives, the gorilla. Similarly, when Strassmann conducted an analysis of the mating strategies of all the primates and compared it to our best understanding of which of these animals menstruate, she did not find a link between menstruation and promiscuity. Those primate species that have the widest variety of mates and are exposed to the widest variety of sperm—those that most need menstruation under Profet's hypothesis—are not menstruating the most.

Many scientists have derided Profet's idea as obviously absurd. How silly that some woman tried to argue that men are dirty! How appallingly feminist to try to turn the scientific tables! Never mind that Strassmann's counterargument was careful, coherent, and objective: the interpretation has been to distort the theorizing of two esteemed scientists into a lady fight. Even as people enjoyed Strassmann's counterargument to Profet's sperm-borne pathogens, few seemed to accept the hypothesis that Strassmann herself put forward for the evolution of menstruation: that of energy economy. She proposed that menstrual cycles evolved because of the energetic cost of indefinitely holding onto a heavily remodeled endometrium.[38]

The central idea from Strassmann is that once the uterus spends the first half or so of the cycle growing an endometrium and at least a week or so remodeling into something suitable for pregnancy, it faces two options: to maintain that nice, rich endometrium with all of its nooks and crannies and lovely growth factors suitable for joining with an embryo or, if that looks unlikely, to scrap it and make it again with the next ovulation. Strassmann did the math and proposed that it is less energetically costly to start over each cycle than to hold on to the same layer of endometrial tissue until a viable embryo comes along.

The need to be economical with one's reproductive system makes some intuitive sense, given much of what we know about growing fetuses and babies. For instance, it has been estimated that pregnancy costs us two hundred to four hundred extra calories a day, while lactation costs four hundred to six hundred. As my colleague Dr. Grazyna Jasienska likes to say, that's a lot of Snickers bars. The question is, are the processes of preparing for pregnancy—growing, remodeling, and maintaining an endometrium—costly enough that the body might have some evolved ways of managing those extra energy needs?

There are a few lines of evidence to support Strassmann's hypothesis. First, it does appear that the more energetic resources a person has, the thicker their endometrium.[39] In my own work, I have found that women with greater iron stores also have thicker endometria, a repudiation of the idea that menstruation within the normal range is associated with iron deficiency. Women with lower progesterone, typically associated with lower energy availability, also tend to have shorter luteal phases (the part of the cycle between ovulation and menses—in other words, these individuals are not maintaining that pregnancy-ready endometrium nearly as long), and I have found that endometrial thickness decreases rather than is maintained in these cycles.[40] Individuals with fewer resources are more likely to dump their endometrium with a shorter luteal phase, have a shorter menstrual phase with less copious menstruation, and/or start over again with a new cycle. (The energetics of the menstrual cycle is of critical importance to how we understand variation in periods and in how we pathologize them; I will deal with this more directly in chapter 3.) Despite the importance of energetics, Strassmann's hypothesis has not been widely accepted.

Instead, a different hypothesis has become the most popular one. Put forward by the biologist Dr. Colin Finn, this hypothesis suggests that menstruation is a simple by-product of the basic biological principle of terminal differentiation. According to this principle, once cells differentiate and specialize to a certain degree, their ability to maintain that level of specialization reduces their longevity. For instance, platelets last hours, T cells last six months, and brain cells last much longer: once differentiated, different cells have different shelf lives dependent on their purpose. The purpose of differentiated endometrial cells is to provide a good environment for pregnancy, and according to this hypothesis, a receptive uterine environment is only optimal for a short period of time.

We do know this from a few lines of evidence: the survival rates of sperm, eggs, and blastocysts, as well as the six-day window during which all embryo-uterine joining occurs. Once that window is up, the terminally differentiated cells are no good, so it makes sense to dump them and start over with a new cycle. Finn argues that the shedding of endometrial cells in the form of menstruation has no particular adaptive purpose or value; endometrial cycling is best explained merely by the straightforward physiological phenomenon that cells die when they need to die.

Among uterus-having species, there are two ways to make an endometrium and initiate the processes of reproduction. There are those species in which different steps, like ovulation or preparation of the endometrium for pregnancy, are induced by an external factor and those species for which these steps happen spontaneously. Induced ovulators, such as camels, llamas, rabbits, and cats, release an egg only when there is an external trigger present like vaginal stimulation from procreative sex, semen, or certain pheromones. In such species, the endometrial tissue does not differentiate to support pregnancy without the presence of an embryo. As you might imagine, this strategy conserves a significant amount of energy: an individual only goes through certain costly reproductive processes if it's worth their investment.

Of course, we humans do it the harder way: we are what's called spontaneous ovulators. This means that we ovulate on our own, without the need for an external trigger. Our endometrial tissue grows and differentiates into something hospitable for an embryo whether or not an embryo is there. For humans and many other mammalian species, the endometrium responds instead to progesterone from the corpus luteum, the "yellow body" left behind after ovulation. The tissue remodeling of our endometrium is tied to ovulation, rather than gamete fusion (a term I prefer over fertilization, for reasons that will become clear in chapter 2). Spontaneous ovulation appears to be wildly wasteful of resources. We do the ovulation, differentiation, menstruation dance all the time, whether or not any of this stuff is going to be used to make a baby. This is part of the reason that Finn's hypothesis is so powerful: there is physiological inevitability behind our being spontaneous cyclers, an ancestral leaving with which we are stuck. Terminal differentiation and menstruation are just what happens when you have cycles determined by internal processes rather than an external trigger.

For decades, Finn's view that menstruation is a nonadaptive byproduct of terminal differentiation has been the prevailing hypothesis regarding why we menstruate. And certainly, terminal differentiation is a phenomenon relevant to endometrial cycling and likely important to the short period during which the endometrium is receptive to the embryo. There are, however, a few things that increasingly do not sit right with me about this theory.

First, the layer of endometrial cells that is shed is heterogeneous: it contains many different types of cells that all have different shelf lives. It's unlikely that they all expire exactly, say, twelve days after ovulation. Second, and I'll return to this in chapter 2, it may not be accurate to call the differentiation of endometrial cells "terminal." In pregnancy, these cells continue to remodel and feed the blastocyst for weeks as the placenta is formed. While there is a period when endometrial cells are optimal for initiating pregnancy, they do not exactly have an expiration date.

The third issue I have with Finn's view is that, to my mind, calling menstruation a by-product of endometrial cycling implies that menstruation is itself biologically meaningless. To Profet, we menstruate to expel sperm-borne pathogens; to Strassmann, we menstruate to cut losses and conserve energy. To Finn, we menstruate to expel the additional blood and tissue of our thicker endometria, but in and of itself, menstruation has no purpose. For the uninitiated, this is about the worst burn in all of evolutionary biology: that one's beloved trait under study is not adaptive but simply along for the evolutionary ride. Useless, rather than useful. Accidental, rather than selected.

It is worth interrogating these moments when characteristics most commonly associated with femininity are rendered useless. Many of my colleagues would hasten to point out that there is no value judgment in the claim that menstruation is not adaptive and that to say so is itself a type of fallacy. I would argue that because scientists are as steeped in culture as any other human, we have a particular responsibility to interrogate these moments to find our own biases. I have observed decades of blowback to functional and feminine-centered interpretations of breasts, clitorises, orgasms, and even grandmothering among the anthropological patriarchy. Given anthropology's history of removing women from the hero myths of human evolution or never noticing their worth in the first place, menstruation is worth a closer look.

As it happens, there has been more recent work to understand the evolution of menstruation and to integrate processes of menstruation into our broader understanding of processes of endometrial cycling.

WHAT IF PERIODS HAVE PURPOSE?

There are good evolutionary reasons for redundancy to develop in biological systems, and something as important as reproduction should have many chances attached to it. But why do we have so very many menstrual cycles and so very many menstrual periods when almost none of them result in conception, let alone pregnancy or live birth? We menstruate a *lot*. Those of us living in industrialized countries are estimated to experience upward of four hundred menstrual cycles over the course of our lives, and there are many examples in both the written record and oral tradition that refer to menstruation as occurring monthly. The very word "menstruation" derives from the Latin *menstruus*, which means monthly, an origin that is shared across most of the major European languages (for example, in Poland you could use *menstruacja*, but you could also use *miesiączka*, which is a diminutization of *miesiąc*, or month).

Periods are less common in our early evolutionary history— monthly-ish at times but with longer intervals of absence. One reason for this has to do with the foraging lifestyle of our ancestors and how that lifestyle may have involved some energetic constraint. This type of subsistence behavior is somewhat hand-to-mouth: humans who live in small groups and move with the seasons have fewer opportunities for food storage and so will often eat what they collect. Foraging encompasses a much wider range of behaviors than the small, nomadic group hunting large game on the savanna, and not all early human foragers were living in scarcity. Among modern foragers, anthropologists have shown they have plenty of leisure time and do not expend any more energy most days than supposedly lazy Americans.[41] But American energy intake is likely far higher, to the point where we typically meet or more often exceed our energetic needs most days. This energy availability makes all the difference.

The other reason we often assume periods were infrequent in our history is due to what's called natural fertility. Natural fertility refers to populations that do not have access to modern contraceptives and therefore have more limited abilities to control the timing of their births. This claim is complicated by the fact that we have substantial evidence of the early use of herbal abortifacients, as well as the presence of postbirth sex taboos and a stigma toward births that were spaced too close to one another. People have known where babies come from for a long time, and it's hard to imagine that birthing parents didn't use all sorts of means to limit procreative intercourse when they wanted to avoid pregnancy. That said, without access to effective contraception, people in natural fertility environments tend to have more children, and in many if not most of these environments, breastfeeding for years is the norm. As a result, people in such environments can go a year or more without menstruating many times over the course of their lives.

While we can be certain that these factors significantly reduced the number of periods our ancestors experienced over the course of their lifetime, it is more difficult to estimate the size of this reduction. Apart from a single conference abstract, I don't know of any studies estimating the lifetime number of menstrual cycles in natural-fertility forager women.[42] However, Strassmann has calculated the number of menstrual cycles in the Dogon of Mali, a natural-fertility farming population. While the Dogon still use menstrual seclusion practices, which make it easier to obtain a reliable measure of menstrual frequency, Strassmann was incredibly careful with her research. Rather than take it as a given that all people visiting the menstrual huts were menstruating, she instead relied on hormonal data to confirm their status. Using the data she collected over two years, as well as information about Dogon women's median age at menarche (sixteen) and menopause (fifty), she estimated that, on average, Dogon women have 128 menstrual periods during their lives. Due to the timing of childbearing and lactating, these 128 cycles are not evenly dispersed throughout a woman's lifetime. Rather, more frequent cycles—eleven to eighteen over the two-year study—were concentrated in both the early part of the reproductive span (fifteen to nineteen years of age) and the later part (thirty-five years to menopause). The years with the greatest quantity of childbearing, twenty to thirty-four years of age, were those in which women experienced very few cycles (three to six over a two-year period).[43] So while this shows there are many years of pregnancy and lactation during which periods occur in clusters, there are still decades of a pastoralist's life near menarche and menopause when frequent periods are the norm.

Compare what we have learned about the forager or farmer cycle to the menstrual cycle experiences documented in many industrialized populations. The latter begin much earlier in life—at around twelve and a half years. Regular cycling continues for at least fourteen years: the current average age at first birth for women in the United States is about twenty-six and a half. Most American cycles resume within a year or so after childbirth, if not much sooner, and go on for far longer before perhaps having another kid. Then it's more cycles, say, from between the ages of thirty-two and fifty. Those who are child-free can expect no particular interruptions in their periods. Of course, hormonal contraception, particularly long-acting contraception, can change the frequency of periods as well. But modern studies of the cycle, by virtue of their recruitment practices, often exclude people who are not cycling regularly or who are using contraception, and they frequently exclude those who are not cisgender women (people whose assigned sex/gender does not align with their actual gender).

Yet despite the significant differences between the cycles among people with different subsistence patterns, both groups still experience many more periods of menstruation than pregnancy. This mismatch becomes even more striking when we consider that an ovulatory menstrual cycle, should procreative sex be timed appropriately, has about a 40 percent chance of leading to conception. One study that sampled healthy Norwegian women showed they only ovulate about two-thirds of the time they menstruate.[44] This means many people seeking to get pregnant (or at least not trying to prevent it) are going to take several cycles for ovulation, timing of procreative sex, endometrial receptivity, gamete fusion, embryo-uterine communication, and the like to lead to pregnancy. And given that humans have a high rate of miscarriage, only half of those pregnancies, if not fewer, will result in a live birth.

Is it possible, then, that periods and all the endometrial remodeling and repair that accompany them might be good practice? This is the rationale of a new generation of scientists who have put forward the idea of menstrual priming.[45] As they see it, processes of endometrial remodeling (terminal differentiation) and repair (menstruation) help teach the uterus how to grow a great site for the embryo and how to foster good invasion of the trophoblast (a layer of tissue outside the embryo that provides nourishment but also helps form the placenta). Menstrual blood is itself crucial to these processes. The idea that endometrial cycling, menstruation, and menstrual blood evolved for this purpose has two lines of evidence. First, as the following deep dive into menstrual effluent will make clear, it is in fact integral to endometrial cycling, rather than a waste product. Second, the frequency of menstruation and certain conditions that arise among people who menstruate at a lower frequency support the idea that frequent menstruation is important to health and reproduction.

To make the first point, which is that menses itself plays a biologically meaningful role in endometrial cycling, let's begin by figuring out what is in there. Menstrual effluent is blood and endometrial tissue. Specifically, it's the blood secreted from the spiral arteries in sufficient quantity to help push the upper two-thirds of the endometrium—the functionalis—off and out. In an ovulatory menstrual cycle, if gamete fusion doesn't occur, the corpus luteum (the yellow body left behind in the ovary from the ovulated follicle and responsible for the majority of progesterone production) starts to degrade. As a result, its progesterone production declines. This triggers an inflammatory response in the endometrium much like the ones you would find elsewhere in the body. Inflammatory cytokines and other immune cells rush in to start doing their jobs, and in particular, enzymes called matrix metalloproteinases show up to help break tissue down.[46] Eventually, the endometrial tissue and fluid starts shedding, and if you're a menstruating person, you find some on your toilet paper, your underwear, or, if you're unlucky, your conference room chair in the middle of an important meeting (don't ask me how I know).

Yet menstrual effluent is not *just* blood and tissue. It also comprises all sorts of cells, hormones, and biomarkers that are involved with the processes of endometrial inflammation and repair. One of the more recent findings is that menstrual effluent contains mesenchymal stem cells, or MSCs. MSCs were first identified in the early days of bone marrow research because they were the first cells to stick to a microscope slide. Yet they've been tricky to fully characterize because MSCs are found in many tissues and have been found to have slightly different properties depending on their origin tissue.[47] In general they show progenitor-like behavior, which means that they can turn into multiple differentiated cell lines and that at least some portion of them are able to self-renew, or make more of themselves.

Endometrial MSCs are crucial to regeneration of the endometrium after menstruation. They differentiate into decidual cells—which play an important role in very early pregnancy—and they are involved in endometrial repair in the three to five days after menstruation has started. MSCs hang out on the stumps of endometrial glands, orchestrating the processes of repair. You can measure MSCs in menstrual effluent, which, remember, is not only making its way out of a menstruating person's body but also hanging out in the uterine cavity. In addition to stem cells, menstrual effluent also contains proteins that aid in repair processes. In 2019, a team of researchers from Australia examined the profiles of proteins in both menstrual effluent and peripheral venous blood, finding almost two hundred proteins that were more abundant in menstrual effluent than peripheral blood and another eighty-four proteins exclusive to menstrual effluent.[48] They found enzymes and enzyme inhibitors that likely work together to produce optimal levels of breakdown and repair. And they found elevated antimicrobials as well as antioxidant enzymes. Therefore, as it is leaving the body, menstrual effluent is a necessary component of the environment in which endometrial repair occurs so that proliferation and differentiation can happen again.

Tissue remodeling during and after menstruation is a remarkable thing. Think back on times you have sustained even a small injury—slicing your finger while cutting vegetables or skinning your knee after slipping on an icy run. In each of these cases, the inflammatory process and repair takes days and usually leaves you with a scar, even if it fades with time. The entire inside of the uterus undergoes a similar process of repair, over and over again, without the buildup of any scar tissue.

Recognizing these extraordinary properties, biomedical engineers are even starting to use the study of the endometrium and menstrual effluent to inform research and treatment for other conditions, ranging from wound repair to cardiac health. In the Australian study mentioned above, for instance, the researchers applied menstrual effluent to wound-healing models both inside and outside the body—in both cases the application of menstrual effluent improved repair mechanisms. While I would not recommend turning a used tampon into a poultice (recall that confirmatory studies of the menotoxin were likely related to uncontrolled infections in blood), it seems increasingly clear that menstrual blood has important healing and repair properties while it is in the uterus. Given the crucial role that menstruation plays in endometrial cycling, we are building evidence against it being a *nonadaptive consequence.*

The second line of evidence in support of menstrual priming relates to preeclampsia. Preeclampsia is a hypertensive (meaning related to blood pressure) condition in pregnancy that can be dangerous and even deadly for mother and fetus. Preeclampsia is believed to be caused at least in part by the shallow formation of the placenta in the uterus and poorly formed spiral arteries compared to a non-preeclamptic pregnancy. Early signs of preeclampsia can include elevated blood pressure and protein in the urine. As preeclampsia progresses, however, mothers can experience swelling in the hands and feet, vision changes, severe headaches, and nausea. Left untreated, preeclampsia can lead to HELLP (hemolysis, elevated liver enzymes, and low platelet count) or eclampsia, preeclampsia with seizures, and death. The only cure for preeclampsia is giving birth, and even after giving birth, recovery can take a long time and require significant treatment.

Preeclampsia occurs most often in teen pregnancies and in first-time pregnancies. It also occurs more frequently in birthing parents who had less exposure to the genetic father's sperm, either because the baby was conceived after long-term use of a barrier method of contraception or because the pregnancy was the result of a shorter sexual relationship.

Preeclampsia also occurs more often in very thin and very heavy birth-ing parents, among those with already-high blood pressure, and, at least in the United States, among nonwhite people. While preeclampsia is thus a risk that runs along many different life paths—bad luck, genetic predisposition, hypertension, psychosocial stressors, early childbear-ing, and contraceptive and sexual choices—advocates for the theory of menstrual priming would contend that at least some of these paths are treacherous because they point to a history of less endometrial-cycling practice. This would be the case for young pregnancies (fewer total menstrual periods) and first pregnancies (fewer established connec-tions between embryo and uterus). The latter idea receives further sup-port from the fact that people who have had an abortion have a *reduced* risk of preeclampsia in future pregnancies. Moreover, these particular paths have evolutionary significance since a disease that may kill a birth-ing parent on the first try reduces their potential reproductive success to zero.

I was filled with a cautious delight when I first found this research. To push against an idea as entrenched as terminal differentiation, though, I scrutinized the evidence with appropriate care. My first question was: What is the incidence of preeclampsia among nonindustrialized populations? For many of these populations, the norm is similar to the Dogon: people start their periods later and give birth earlier. In gen-eral, high-fertility and natural-fertility populations menstruate less, which, according to the theory of menstrual priming, should put them at increased risk for preeclampsia. If we saw a very low incidence of preeclampsia among such populations, it would constitute a major hole in the argument.

The answer to this question is not straightforward. Only one study compares the prevalence of preeclampsia across the globe, and it found that women from African samples had the highest prevalence of pre-eclampsia.[49] Turkey, Nigeria, South Africa, Argentina, Slovakia, and Indonesia have higher rates of preeclampsia, though few of the studies described how they defined or diagnosed preeclampsia. Looking at the United Nation's data for average age at first birth among the popula-tions, I found that none of the countries with the highest incidences of preeclampsia had especially low ages at first birth, though most were lower than the average in the United States (which is about twenty-six years). Perhaps we see more preeclampsia in populations that are less industrialized and/or have a younger age at first birth.

Next, I looked at the incidence of preeclampsia among immigrant first-time mothers in industrialized countries.[50] Immigrant women from sub-Saharan Africa, Latin America, and the Caribbean had higher rates of preeclampsia than those from Western Europe, even after controlling for age. Based on these data, it's safe to say that those living in less industrialized conditions, and potentially even more rural conditions, do have greater risk of preeclampsia. However, remember that there are multiple paths to preeclampsia, including environmental factors like psychosocial stress: it is well established that population health disparities that appear to be about "race" are in fact about racism.[51] It's not possible to know for certain the extent to which the variation we can see in preeclampsia across different environments is from less endometrial practice, more stress, or something else entirely.

An additional line of evidence that would help us evaluate the theory of menstrual priming concerns hormonal contraceptives: Do we see more preeclampsia in those who use hormonal contraceptives and therefore do not go through quite the same process of significant endometrial remodeling before menstruation? Again, there isn't much in the way of direct evidence. One study looked at birthing parents who had used barrier methods, nonbarrier methods, or no contraception before getting pregnant and whether they had developed any hypertensive disorders (of which preeclampsia is one). This study found that there was a slightly increased risk for hypertensive disorders among the women who had used nonbarrier methods—presumably those that were hormonal in nature.[52] These would be the contraceptives that either lead to less endometrial remodeling (and lighter periods, like the pill) or none at all (no periods, like hormonal intrauterine devices, or IUDs). However, the relationship between hormonal contraceptives and hypertension in pregnancy might also be related to the fact that they can increase blood pressure in their users.

THERE IS A REASON FOR ALL OF THIS   49

There is one final piece of indirect evidence for menstrual priming, which I found in a paper on the menstrual bleeding patterns of highland Bolivians. Anthropologist Dr. Virginia Vitzthum and her colleagues have studied the menstrual cycles of Bolivian women for decades. In a 2001 publication focusing largely on the experiences of menstrual bleeding among lactating versus nonlactating women, Vitzthum makes an important finding: women who conceive often have longer periods in the previous menstrual cycle.[53] This could be an indicator of energetics: women who have enough energy in the tank to have a thicker endometrium will have a longer menstruation. But it might also be the case that those women who conceive may be women who have had more immediate and thorough endometrial-cycling practice.

Where does that leave us? Is menstruation toxic? Is it economical? Is it useless yet dragged along by the necessity of our terminally differentiating endometrium? Or does menstruation help prepare the endometrium for later pregnancy? We should not be so quick to assume that there is only one right answer. By imagining the question as multiple choice, we risk assuming that our bodies are simpler than they might be. In evolutionary biology it is often tempting to choose what's called a *prime mover hypothesis* for a given trait—to claim that there is one main reason that something evolved. In human evolution some have said that language is the main reason we grew big brains, while others claim that we developed big brains to foster social relationships and still others that it was to coordinate hunting or find challenging food sources. As Dr. Mary Rogers-LaVanne and I have recently argued, the problem with applying prime movers to menstruation (and really any other biological process or trait) is that the idea that there can be only one right answer is an especially white, Western, colonial one. We come from scientific traditions of discovery and pioneering, in which credit is often attributed to one person. Much of our culture, including the origins of Western science, promotes and rewards hierarchies of knowledge.[54] Maybe there just isn't a hierarchy. It is more realistic to say that the menstrual period is a physiological inevitability because of how endometrial cycles work, that menses is not somehow separable from that process, and that surely saving a little energy here or there in the process helped make it what it is today. We can all play in the same sandbox and still be friends.

Now that I've delved into the often-neglected question of *why* menstruation happens, I'd like to explore the how. How do menstrual cycles and menstruation work? How do the sensations and experiences of the menstruating person match up with what is happening on the inside? Finally, how have our most basic assumptions about what constitutes "normal" and "healthy" led past researchers astray? We've seen how white, Western, and/or colonial perspectives have colored scientific interpretations of why we menstruate; it's eugenics that has complicated the how. Trying to separate eugenics from anthropology is about as easy as unscrambling eggs. However, our efforts to notice where a eugenic bias has informed our understanding of menstrual cycles have paid off in the new types of analyses my lab has been able to perform to address the question of what constitutes a "normal" menstrual cycle.

CHAPTER 2

Norma's "Normal" Cycle

IN 1945, the American Museum of Natural History revealed two new statues, Norma and Normman. These statues did not, however, seek to depict any specific persons. Instead, they were composites, meant to represent the "statistically average" American man and woman. A collaboration between gynecologist Dr. Robert L. Dickinson and sculptor Abram Belskie, the statues drew from a variety of sources: a few million World War I soldiers, "special studies of the old American stock," college women and men, a sampling of attendees at the World's Fair in Chicago, insurance records, and a sample of fifteen thousand white women used by the Bureau of Home Economics to determine sizing standards for women's clothing.[1] Norma reflected the norm for an eighteen-to-twenty-year-old American woman while Normman was the norm for a twenty-year-old man.

These statues reaffirmed the ideal of the able-bodied, white, straight American and brought together the various paradoxes of what it meant to be normal in the twentieth-century United States.[2] "Normal" meant average, but not across all populations. Instead, normal was calculated across only a certain set of desirable populations, excluding people of color, queer people, disabled people, and anyone who might dare to occupy multiple of these identities. In fact, Norma and Normman were expressly compared against North American Indigenous people to quell white anxiety that the American environment would, over time, degrade good European stock. Dickinson, who pitched Norma as both "the perfect woman" and "the average American," was influenced in his work by the eugenicist and anthropologist Earnest Hooton, who advocated against aiding the poor and the "reckless breeding of the unfit."[3]

American Museum of Natural History curator Harry Shapiro, also a eugenicist and anthropologist (unfortunately, these roles often went hand in hand in the mid-twentieth century), wrote in a companion piece to these statues that "the existence of a physical type, distinctively American, has for a long time been quite generally accepted both by Americans themselves as well as by foreigners." In this way Norma and Normman were meant to represent typical (white) Americans, distinguishing them from Europeans, particularly the English and the French. Shapiro, who would go on to become the president of the American Eugenics Society ten years later, found the physical features of these American averages quite pleasing, remarking that they "leave little doubt that the figure is improving esthetically."

Of the two statues, it was Norma that captivated the public's interest. When Norma and Normman were moved to the Cleveland Health Museum, the museum, along with the Academy of Medicine of Cleveland, the School of Medicine, and the Cleveland Board of Education, sponsored a contest to find a woman who most closely matched Norma's dimensions.[4] The Ohio woman who won would get a one-hundred-dollar war bond and earn the title of "Norma, Typical Woman." Of the 3,864 contestants, only 40 hit the right dimensions on five of the nine categories; even the winner, a twenty-three-year-old theater cashier named Martha Skidmore, did not achieve normality in all nine dimensions. As Shapiro himself admitted, gorgeous as these forms are, they do not seem to represent many living people: "Let us state it this way: the average American figure approaches a kind of perfection of bodily form and proportion; the average is excessively rare."

Nowadays being normal is often an enforced set of traits, a way of ingrouping and out-grouping people. As Drs. Peter Cryle and Elizabeth Stephens, scholars at the Institute for Advanced Studies in the Humanities at the University of Queensland put it in their book *Normality: A Critical Genealogy*, "The word 'normal' often suggests something more than simply conformity to a standard or type: it also implies what is

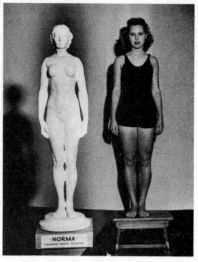

The statues of Norma and Normman.

Norma with prizewinner "Normal Woman" Martha Skidmore.

correct or good, something so perfect in its exemplarity that it constitutes an ideal." To be outside the range of normal is to be abnormal—pathological, queer, or an ill fit within a certain set of culturally prescribed categories. This definition of normal has a specifically American and eugenicist origin. As Shapiro argued regarding Norma and Normman, normal manages to be both ideal and represent the average, in that case rendering the American form more superior and evolved in his comparison to Europeans. While in the nineteenth century normal meant something closer to function, or being in a state of equilibrium, as we headed into the twentieth century it came to be understood as an outcome that one could quantify. For example, when you get blood work at the doctor's office, you are told your own values alongside the "normal range."

This framework is also how many of us are taught about menstrual cycles. There is an expected and "normal" pattern to our hormones and the responses of our tissues and organs. In the first half of the cycle, known as the follicular phase, estrogen goes up and along with it a thickening of the uterine lining. An egg is released precisely at midcycle, and then progesterone sticks around for a while in case there is sperm-egg fusion and the beginnings of pregnancy. If not, progesterone goes down again, and the uterine lining is shed. It is a tidy and predictable four-week experience found in a wide variety of puberty books, health classes, and medical texts. In short, we've all been taught some version of the menstrual cycle that belongs to Norma and Norma alone.

In this chapter, I introduce the idea that if we truly want to understand menstruation, it's better to seek to understand processes rather than outcomes and to explore variation rather than dichotomize people as normal or pathological. The processes of the menstrual cycle are active, and knowing these processes will help us understand the many points at which we depart from Norma's menstrual cycle—departures because of immaturity or aging, because of stressors, and because of lifestyle. First, we'll walk through the processes of puberty, as one hits menarche, or first menstruation. Next, we will linger a bit with the ovaries, discussing where they fit in the broader processes of reproduction, how ovulation works, and how eggs are selected, as well as how the ovaries and the uterus have been badly misapprehended as passive observers of reproduction. Finally, we will return to this idea of normality and ask: What if all of these different patterns, of first periods and menstrual cycles and hormones, are normal after all?

FIRST PERIODS: THE PHYSIOLOGY OF MENARCHE

I got my first period the very first month of eighth grade, September of 1992. Because it was associated with school starting and thus was pretty memorable, I can confidently say I hit menarche at thirteen years and two months. I remember my mom's smile in that early morning moment, indicative of the way the adults around me treated it as a major transition. Yet I had been feeling increasingly different for years. I remember being bullied by the teammates of my community basketball team in sixth grade because I didn't yet shave my legs, then getting yelled at by my ballet instructor in seventh grade because I didn't yet shave my underarms. There is a photograph from my ballet recital in seventh grade, which was placed in a prized position on my parents' piano, that I refer to as my "mortifying breast bud picture" because they were so prominent through the spandex of our early nineties costumes. The arrival of my first period was just one step along the very long, tortuous path of adolescence.

Many years of physiological development—of the brain, multiple organs, and the hormonal messengers that keep everything running— precede that moment when a young person wakes to a stain on their sheets. One of the first major developmental events is called adrenarche, which is characterized by an increase in androgens (specifically, ones called DHEA and DHEAS) from the adrenal glands that precede the production of other hormones like testosterone and estrogen. The adrenals are small triangular organs that sit on your kidneys. They have several functions, but the most pertinent here is the production of glucocorticoids like cortisol, which help regulate many systems, including reproduction, and the eventual production of other androgens from the initial increase in DHEA and DHEAS.

Adrenarche is more of a process than a defined event—as the adrenals slowly increase the production of DHEA and DHEAS above a certain medically defined threshold—and it occurs approximately between the ages of six to eight years. The next stage, for most people with uteruses, is thelarche. Thelarche is the development of breast tissue, and whereas adrenarche has one well-defined window, the timing of breast development varies greatly between people. A recent study, for instance, showed considerable variation in thelarche by migration status. Among white British girls born and raised in Great Britain, thelarche began at about eight and a half years of age, give or take a year, while British-Bangladeshi girls who were born in Bangladesh but raised in Great Britain initiated thelarche about a year later, and girls born and raised in Bangladesh reached thelarche more than two years later.[5] This variation is likely a function of the girls' different living conditions at crucial developmental points.

After thelarche is pubarche, which is the development of pubic and other axillary hair, like in the armpits. And after that, at least for most children with uteruses, is menarche, or that first period. Along with these increasingly outer displays of pubertal development, there is a rich broth of hormonal messengers being cooked up in the body. Certain parts of the brain wake up, ready to contribute to the signaling system of the ovaries. The entire reproductive system is maturing, perhaps even technically ready to reproduce before one has their first menses. In one study that measured salivary hormones in girls in the months leading up to menarche, five of the forty-two were shown to produce enough progesterone that they had likely already ovulated.[6]

And the processes of puberty don't end with the first use of a pad, tampon, or cup. Rather, the relationship among the brain, the ovaries, and the adrenals continues to mature for years. In the first one to seven years after menarche, it is typical, even desirable, to have irregular periods. Less frequent periods during this time result from being exposed to less estrogen, and the desire that many have to regulate periods this early in life does not reflect the reality of how bodies develop. The idea that a normal menstrual cycle, from menarche to menopause, is supposed to be exactly twenty-eight days long seems to have more to do with a desire to keep things simple and tidy and Norma-like than to explore the different functionalities of the menstrual cycle across the life span.

Menarche appears to outwardly signal the major accomplishment of puberty, the production of menses. However, it is perhaps better thought of as a halfway point between the initial development of hormonal signaling pathways and the transition from a body that grows to a body that can reproduce. In one study in the United States, women were interviewed about their menarcheal experiences. Most participants self-classified their menarche as "early" or "late," but the ranges reported by these two groups were right next to each other—nine to eleven and a half and twelve to seventeen years, respectively.[7] Almost no participants felt they had gotten their period "on time," and many expressed significant embarrassment about the timing of their menarche.

The medical literature about menstruation also focuses quite a bit on the problems of early or late menarche and espouses a general worry about the decline in menarcheal age across many human populations for some time.[8] Among industrialized populations, an average age at menarche is about twelve and a half years, give or take a year. Older studies that collected data from modern foragers estimate an average age of menarche as late as sixteen or seventeen, which is often assumed to be close to our ancestral norm.[9] A lot of my research over the last twenty years has taken place in rural southern Poland, among farming populations where people still do a lot of their farming by hand or only with the aid of horses. They have large gardens and until recently supplemented little of their diet with food from their local *sklep*, tiny shops that carry a little produce, alcohol, ice cream, and other sundries. The women we have sampled in rural Poland have a slightly older average age of menarche than what we see among more industrialized populations but nothing like the estimates for foragers.[10] According to a paper out of our lab led by Dr. Mary Rogers-LaVanne, using several decades of data across several collaborators, the age at menarche has declined quite a bit in this region. Women reaching menarche in the 1960s were on average fourteen and a half; by the 1990s, the average age at menarche for this population was a full year younger, about thirteen and a half.[11] Of course, now many farms have been bought up by corporate farming entities, and instead of having to do all of their own gardening for vegetables, these villages now have more access to large grocery stores with produce and other food items from all over the European Union.

Overall, the decline in age at menarche across many human populations mostly reflects what we see in Poland: a shift in energy balance from an increase in food availability, a decrease in physical activity, or both (as we will discuss in more detail in chapter 3). Young people with a positive energy balance can allocate more resources not only to pubertal development but also to skeletal growth, which means that they don't need to stretch out puberty to reserve energy for growth. Within a given population, however, there remains substantial variation in age at menarche that cannot be explained so simply.

In the early days of research on menarcheal variation, many scholars focused only on the negative: they contended that factors like father absence, stepfather presence, adverse childhood experiences, or other stressful factors contributed to early menarche. Among US populations these correlations did often (but not always) bear out, and researchers hypothesized that individuals living in uncertain environments reach reproductive maturity more quickly in order to initiate reproduction.[12] If you are unsure how long you will live, it might make Darwinian sense, at least, to try to have some babies sooner. More recently, work from our lab and others has pushed back on this idea that those who reach menarche early somehow have lived a harder life than those who start menstruating later.

One problem with this earlier research is that, among non-US populations, father absence is not always associated with earlier menarche, and the reason for the absent father—from death, divorce, or migrant work, for instance—significantly changes its effect. Indeed, a recent paper with a more global view has shown, particularly among populations that are less well nourished, that the absent-father hypothesis is not well supported.[13] First, it's important to note that about 50 percent of menarcheal age is heritable—that means that about half of the variation in age at menarche can be explained by when one's relatives reached menarche. But the larger problem, in my view, is that this research does not consider how positive childhood experiences, with strong parental relationships, might influence menarche. My lab sought to address this deficit during the six summers we collaborated with psychologist Dr. Carla Hunter and bioengineer Dr. Jenny Amos to work with high school–aged girls attending a STEM (science, technology, engineering, and mathematics) camp in the United States. Our educational goals were to expose them to the kinds of social science education that would help them handle the inequities that come with being a woman in STEM, as well as make them better scientists and engineers by teaching them that any applications they work on, from nanosensors to prosthetics, would need to work on a range of bodies and not just Norma and Normman.

But we did have these biological research goals too. In particular, we were curious about the role of social support on menarche: the menarcheal period involves not only significant physiological change but psychological change, with a transition away from parental support and toward peer support. To study this, we asked students to fill out a social network inventory, which in our case meant they were asked to list the names, relationships, and strength of connection of the top twenty people in their lives. Some girls ran out of space and added lines to put on more names, while others struggled to come up with twenty. Some girls asked if they could add their pets, and so our data set includes some cats, dogs, and even a chicken.

We then asked the girls to create social network artwork that visually represented this list. Putting their names in the center of the sheet, some girls drew trees, with the proximity of branches and their thicknesses representing closeness and frequency of contact; others made spiderwebs; still others, self-portraits or abstract art. We used this exercise as a way for the girls to think about the support they had as they realized their dream of becoming scientists and engineers. Even the oldest and grumbliest of the lot usually ended up enjoying this exercise and the opportunity it gave them to reflect on the people around them.

We then used the social network inventories and a validated scale to assess parent-adolescent relationships to explore how "tending and befriending" might operate differently when there is a loosening of priority with parental relationships. *Tend and befriend* refers to the hypothesis, first put forward by the psychologist Dr. Shelley Taylor, that stressful situations often cause individuals of many species to seek social support and, in particular, that these behaviors are often seen in females through nurturing behaviors (tending) and affiliation with peers (befriending). Dr. Michelle Rodrigues, a former postdoc in my lab who is now a professor of anthropology at Marquette, is an expert on the tend-and-befriend hypothesis, having studied it across several primate species in field and captive conditions.[14] In our sample of ninety-five girls who had attended these summer STEM camps, Michelle found that the strength of the girls' relationships with their mothers predicted the number of female friends they had and that those with more female friends displayed fewer depressive symptoms. Father relationship quality had less of an effect on friendship quantity or quality but did correspond directly to fewer depressive symptoms.

Following up on this research, Dr. Mary Rogers-LaVanne was curious about how parent-adolescent relationships might influence the timing of age at menarche. Mary found that better relationships with mothers corresponded to an earlier age at menarche and better relationships with fathers, a later age. This is like an inverted test of the absent-father hypothesis. Perhaps accelerated development has more to do with having a secure and loving attachment with the person who is most likely to help you out if you start having children and less to do with an "absent father." And perhaps a secure and loving attachment with a father signals a stable environment, where it is safe to delay puberty just a bit in order to continue to focus on skeletal growth and maturation.

People who get their periods early or late are not abnormal, and they aren't necessarily damaged or traumatized. The timing of that first period is simply one indicator of development, which occurs amid a profusion of changes in the body. Of course, now that you know the period itself is just one part of a vast system involving the brain, adrenals, uterus, ovaries, mammary tissue, and more, we should turn our attention to getting to know the latest research on how the rest of that system works.

MESSENGERS AND FEEDBACK LOOPS

The menstrual cycle is the time between the first day of one menstrual period to the first day of the next one. It encompasses a number of physiological processes that are designed to prepare the body for possible gamete fusion (gametes are the general term for sperm and eggs), interfacing with an embryo (this is when the embryo and uterus start talking and finding a place to meet), and pregnancy (the establishment of this relationship between embryo and uterus where they join and uterine remodeling begins to support an eventual fetus).

The processes of the menstrual cycle are regulated by the hypothalamic-pituitary-ovarian (HPO) axis. Because this axis operates through the hypothalamus (a region of the brain), which regulates practically everything, understanding how it works helps us understand how the menstrual cycle is not a closed system; it is instead highly impressible by many factors. As we will see, the menstrual cycle is affected by development, lived experience, age, and more. In the same way Martha Skidmore was unable to fully embody Norma, the normal menstrual cycle represents an ideal so outside of common experience as to be nearly unachievable.

The HPO axis is a big feedback loop: some hormones send a positive message, telling the next gland or organ to make something, while other hormones send a negative message, telling them to stop. The hypothalamus produces gonadotropin-releasing hormone, or GnRH. This exerts a positive effect on the pituitary (a gland in the brain), telling it to make (you guessed it) gonadotropins—specifically, to make follicle-stimulating hormone and luteinizing hormone (FSH and LH, respectively), both of which affect ovarian function. When FSH and LH reach the ovaries, they stimulate processes that help make selection of a dominant follicle and ovulation happen. They also stimulate the production of estradiol. Estradiol exerts a negative effect on the hypothalamus, telling it to stop making GnRH. Within the ovary an overlapping feedback loop is also operating: estradiol and LH provide ever-increasing positive feedback to each other as the dominant follicle is maturing. Progesterone is only produced in significant amounts after ovulation, when the yellow body, or corpus luteum, is left behind in the hope of maintaining a pregnancy. Progesterone is what provides negative feedback to the pituitary to shut down that mini estradiol-LH loop and exerts all sorts of interesting effects on other reproductive tissues, like the uterine lining, or endometrium. Should there be no gamete fusion or other pregnancy-related business, the corpus luteum degrades, progesterone declines, and the process starts again.

In addition to playing a crucial role in the HPO axis, the hypothalamus also regulates a system known as the hypothalamic-pituitary-adrenal axis, or HPA axis. The HPA axis signals the existence of external stressors with which the body needs to deal. As I've mentioned before, the adrenals make glucocorticoids like cortisol, a hormone that rises alongside the experience of many different types of stressors. In this system, the hypothalamus makes corticotropin-releasing hormone, which sends a positive signal to the pituitary, which then makes adrenocorticotropic hormone, which sends a positive signal to the adrenals. The adrenals then make cortisol, which, like the ovaries' production of estradiol, sends a negative signal to the hypothalamus and pituitary to shut off. Because the hypothalamus and pituitary are both involved in these systems and because they are shut off by the production of hormones by the adrenals and ovaries, you can begin to see how the activation of the stress axis, or HPA axis, might also lead to the suppression of the HPO axis. As we'll discover later, cortisol also plays a direct role in some processes in the ovaries and uterus, so there are multiple ways by which stressors, be they from a hard workout, a rude coworker, or a viral infection, can change them.

What I've just described are the processes of the normal menstrual cycle in the original meaning of the term; that is, these are the basic functions of the anatomical system as best as we can understand them. Going forward, the scientific literature is increasingly infected with concepts of normative femininity and heterosexuality in describing ovarian function and how it relates to the uterus and menstruation. Once you start talking about eggs, especially eggs and sperm, it seems people cannot resist equating eggs to princesses and sperm to princes. Or worse: uteruses to baby baskets and sperm to baby makers.

A book by Dr. Rene Almeling, a sociologist at Yale who studies the intersections of gender and medicine, shows that this narrative persists in the minds of laypeople as well.[15] Almeling interviewed forty men about the roles men play in sex and reproduction.[16] When asked, "How would you describe the relationship between the sperm and the egg?," most respondents used the usual active sperm/passive egg narrative. As one respondent explained, "Everybody's just pushing each other out of the way, just to get to the egg, and that's that golden prize, so to speak." Another said, "An egg is not a live breathing thing. It's a cell. Your father's sperm is a live breathing tadpole. It eats. It breathes. It moves. It swims. It's a live breathing organism, whereas your mother's egg, it's a shell. . . . These are just facts. If you really pay attention, you think about biology, the truth, you come from your father." Thankfully, Almeling found that half of the men in her sample eventually told a more equal narrative, amending their stories as they went, such as that they are "two necessary components that need to come together," or "like a 50/50 sharing of information," or that "there is no life without both of those two coming together. So basically, they need each other."

Normative femininity has influenced our understanding of the menstrual cycle in two ways: it has created the idea that the menstrual cycle and its related processes are passive and nurturing, and it has encouraged physicians and patients to feel the need to regulate abnormal cycles toward a normal shape and length. The normal menstrual cycle, constrained by normative femininity, barely resembles the reality of what is going on in our uteruses, ovaries, and other organs and systems.

In the hero's journey, a narrative template famously articulated by Dr. Joseph Campbell, a man goes on an adventure where he faces and eventually overcomes adversity. The heroine's journey, by contrast, has long lacked a clear archetype. When Dr. Maureen Murdock once interviewed Campbell about this subject, he had this to say: "In the whole mythological tradition the woman is *there*. All she has to do is to realize that she's the place that people are trying to get to." You know, like the "golden prize" or "shell" described by some of Almeling's participants. Campbell continued, "When a woman realizes what her wonderful character is, she's not going to get messed up with the notion of being pseudo-male."[17] Because, I suppose, an adventurous spirit is the antithesis of femininity.

Despite the views of Campbell and others, dynamism, strength, leadership, and adventurousness are traits that belong to everyone. Our misappropriation of normative femininity as passive has led to a significant misunderstanding of what happens throughout the menstrual cycle. Rather than a tidy system that ticks along in a prescribed order, the reality of the menstrual cycle is that it comprises a set of processes that are incredibly sensitive to each other and to the outside environment. It's messy and dynamic, so unlike the passive stories often told about the fertile soil of the uterus or the egg waiting for her prince. The person is not just *there*. The menstrual cycle is an agent of its own destiny. No part of this system has more agency than the ovaries themselves.

OVARIES IN CONSTANT MOTION

The ovaries are the organs that make eggs, and if you have ovaries, they already contain all the egg cells, or oocytes, you will ever make (minus any you may have ovulated). A fetus makes all their oocytes by about twenty weeks of gestation—it's a nice safe time to make high-quality gametes because the fetal ovaries are inside a fetus inside a person, protected like the little baby at the center of a set of matryoshka nesting dolls. So by twenty weeks, the birthing parent contains a fetus that contains eggs that could become the birthing parent's future grandkids (go ahead, read it again). It's weird to imagine that the egg that made me once resided in my own grandmother or that the eggs that made my children were once inside my mother. This also means that in times of substantial harm like the Holocaust, the Dutch Hunger Winter, the Rwandan genocide, the COVID-19 pandemic, or even sustained contact with environmental pollutants, three generations can be exposed at once.

Throughout one's life, there are several culling events, also called oocyte atresia, that allow the body to select the oocytes with the best chance of successful gamete fusion, the first of which occurs during that period of fetal development. While the fetus begins with between 5.5 million to 7 million oocytes, by birth they are down to about 1 million; by puberty, only 300,000 oocytes remain.[18] While oocyte atresia continues from puberty to menopause, killing off unworthy oocytes, each menstrual cycle is also like its own culling. The way I learned about eggs and ovulation in health class made this process sound very linear: after a brief selection process among a handful of eggs during the first half of the menstrual cycle, one egg wins the beauty contest. This one egg, or dominant follicle, forces the others to concede the crown and is released from the ovary to rest quietly in the fallopian tube and await her true love.

Research in the past few decades has upended several of the above notions: the process of egg selection doesn't occur only in the days before ovulation. Instead, the ovaries oversee continual, overlapping waves of competition to select a dominant follicle for ovulation. The basic idea behind ovarian follicle waves, which were first documented by Ron Jankowski in 1960, is that follicles develop in coordinated groups. Jankowski cut open the ovaries of cows, performing daily sequencing of follicular growth and development over time. He found that cohorts of follicles grow and die together multiple times between each ovulation. The cow data were, however, cross-sectional—these were postmortem ovaries representing one cow in time, not repetitive sampling from the same cow over time since the only available method to look at the ovary was to pull it out and slice it open. This is akin to testing for the ripeness of a cantaloupe by cutting it in half to look inside. Ultrasound technology, which allows scientists to make real-time images of the reproductive system—to look inside the cantaloupe without cutting it open—wouldn't exist for another few decades.

Fast-forward to the early 1980s when colleague and collaborator Dr. Roger Pierson was starting his PhD at the University of Wisconsin. His adviser procured one of the first real-time ultrasound units in North America, and unlike many units at that time that only worked outside the body, it was modified to have an interrectal probe—yes, that means what you think it means. When I spoke to Roger about it, he declared, as though sharing a warm memory, that the ultrasound unit "worked beautifully with horses because horses have the biggest ovaries out of all domestic animals, and they have the largest dominant follicles."

Over the next few years, Roger identified follicle waves in horses and then cows. As it turns out, he came to learn that "almost all domestic animals have waves of follicular development." He then moved on to rhinoceroses, elephants, and whooping cranes—all of their follicles grew and senesced in waves. These waves occur for the entire duration of the time between ovulations. Once ovulation happens, there is no rest, no holiday, only renewed competition, upending our notion of a passive ovary waiting for its true-love sperm for the whole second half of the cycle.

By this point, most animal science researchers, particularly those who studied domestic animal reproduction, were aware of follicle waves. However, the concept of follicular waves never really took off because the domestic animal research groups were primarily focused on monitoring hormones. Medical researchers never considered the idea that the human ovary might behave in the same way, in part because medical researchers didn't tend to read domestic animal science research. This is where Roger's student and another colleague and collaborator of mine, Dr. Angela Baerwald, came in.

By the time Angie started her PhD in 1997, ultrasound technology and software were well developed: the sustained study of follicle waves was ready for prime time. After taking a class from Roger, she became excited by his early research on follicular waves in other animals and decided to study follicular waves in humans for her PhD work. To conduct this research, Angie performed daily transvaginal ultrasounds from just before one ovulation and all the way through the next ovulation on about one hundred women. Angie, with some help from Roger, was scanning up to a few dozen ovaries every day, seven days a week, for months. Appointments were spaced fifteen minutes apart. Each appointment took about ten minutes, slightly longer on the days when they also collected blood. Angie would do a count of all the follicles in each ovary—the days she had someone in the room with her to write down her counts were the good days—and then also take a video of the ultrasound, slowly scrolling through each ovary. Later, she would analyze these videos, counting and measuring every single follicle in each ovary, a process that took another twenty minutes per scan.

As Roger and Angie had expected, human ovaries aren't all that different from many other mammals: they don't laze about for half of the cycle only to select a follicle in the final days before ovulation, as many health and medical textbooks would suggest. Rather, like so many other mammals, the emergence of a dominant follicle occurs throughout the entire process of ovulation. The process is dynamic and constant—one of several phenomena in the HPO axis that, now that we better understand it, completely defies the standard notion of normative feminine passivity. The ovaries are anything but passive. They are *always* up to something.

The mechanisms that drive follicle emergence during these waves are vigorous and display a level of competition between eggs that rivals the narratives common to flinty sperm. In Angie and Roger's sample of urban Canadian cisgender women, participants had no more than two or three waves of follicle development. And there were two types of waves, major and minor: a *major* wave is one where a dominant follicle emerges—a follicle at least ten millimeters in diameter and at least two millimeters bigger than any of the others in its cohort. Waves without dominant follicles are called *minor* waves. Angie and Roger documented five wave patterns in their sample: minor-major and major-major among the two-wave folks and minor-minor-major, minor-major-major, and major-major-major among the three-wave folks. Note that in these patterns, once someone has a major wave their next wave in that interval is always major, and since the cycles reported on in this sample were all ovulatory, the last wave was always major as well.

The most common wave patterns among the two-wave folks were minor-major, while the three-wave folks most often experienced minor-minor-major. This suggests that, at least among ovulating folks, the ovaries only tend to select one follicle per cycle, despite the constant infighting between ovulations. This process is one main way in which the reproductive system culls eggs in order to create conditions where the best egg wins. However, some number of people—those with major-major and major-major-major patterns—experience multiple surges of FSH and LH that correspond with waves where a dominant follicle emerges but does not ovulate.

Angie and Roger's first papers on ovarian follicular waves in humans are now more than twenty years old. It has taken quite some time for their ideas to gain traction, despite powerful evidence. But thanks to this work, we have better insight into many major questions about the menstrual cycle. For instance, previously, we did not understand why many people near menopause (defined retroactively as the point in midlife at which a person has gone at least a year without menstruating) had not only more irregular but often more frequent periods. The reason those cycles start getting so short for so many people? Changes in their numbers of waves.[19] And while we used to blame hormonal contraceptive failures entirely on patients not taking their pills as directed, we now know that there are two different times that dominant follicles can inadvertently develop: during the hormone-free week that mimics a period and when a person starts taking their pills on Sunday, called the *Sunday start method*, instead of on the first day of their period.[20] Roger and Angie's work has also provided foundational knowledge for newer and better assisted reproductive technologies, both for "poor responders" to hormonal stimulation and for those in urgent need of egg preservation, such as cancer patients.[21] Prior to this work, the protocols to start to grow and mature eggs always began on the first day of the menstrual cycle since this was when doctors thought the ovaries started developing their follicles. Since we now know that this process is more or less continuous, doctors can help folks mature or preserve eggs starting at any point in the menstrual cycle and even make multiple attempts per cycle.

Why did it take so long for scientists to get on board with the concept of follicular waves in humans? Many clinical researchers pushed back on Angie and Roger's work, claiming there was no way they could actually see ovulation. Roger would pull out the videotape and computer files and show it to them in real time with a recorder. Well, they would say, you can't see a new corpus luteum. Pointing to the tape, Roger would show them a follicle losing its fluid, along with the moment the follicle becomes a new corpus luteum. He would show them what the ovaries looked like one day, two days, three days later. At that point, he says, "The silence in the room got pretty loud."

But then I had to ask: Was it at all possible that the cultural beliefs of scientists prevented them from seeing this reproductive process as exciting, competitive, and constantly active? Could that help explain why we missed such an exciting biological phenomenon for so long? Roger replied, "Most of the gynecologists in the world at the time were male and were still looking at reproduction from an androcentric viewpoint. . . . Female biology is an incredibly active process and is probably many, many full times more active than male biological processes." Get a load of *that*, Joe Campbell.

LAST PERIODS: THE PHYSIOLOGY OF PERIMENOPAUSE

Thanks in no small part to Roger and Angie, we now have a much better understanding of the emergence of the dominant follicle. Once that follicle is selected, the next step is ovulation, when the egg leaves the ovary and heads to the fallopian tube. Scientists are still working to fully understand the process by which one follicle beats out the others in the ovarian *Hunger Games*, but we know that the dominant one ultimately produces more estrogen and that these follicles are fighting a game not only of size but of speed. FSH concentrations going down is what causes selection of the dominant follicle, but that decline in FSH is caused by secretions of the other follicles in that wave. This means that for some reason, those follicles are complicit in their own demise.[22]

As a species that mostly ovulates once per cycle and usually just one egg per cycle, we are putting all our eggs into one . . . egg. These cyclic culling events are helping us choose the best egg of each bunch, after multiple rounds of culling have already happened. However, as we age the emphasis placed on quality changes. As the quality of the oocytes declines as they, too, age, the ovaries start loosening restrictions on what constitutes a good egg. This explains why there are some tiny but meaningful increases in genetic abnormalities in the embryos of people over forty, though many of these can also be explained by genetic abnormalities in the sperm of their also-aging partners. This shift toward quantity over quality later in the reproductive span is also why we see multiple eggs per cycle and therefore multiple pregnancies (twins and even triplets) in higher numbers in older parents. This isn't abnormal—this is aging. The function, shape, and timing of processes of the menstrual cycle are vastly different at different points in one's life.

The normal menstrual cycle we have been taught is effectively what we assume to be a common cycle among twenty-five-to-thirty-five-year-olds. But people can menstruate for forty years or more, and cycles in the last decade of this span can be really different from what came before. Perimenopause is the transitional period between the time one is most fertile to the total cessation of menstrual cycles.[23] Menopause is that total cessation of menses. Clinically, it is a moment that can only be defined retroactively: one has to be without a menstrual cycle for twelve months to officially hit menopause. Perimenopause is similarly difficult to define prospectively and frankly is cloaked in far too much mystery. For years, despite the evidence, clinicians often discussed perimenopause as a period of declining estrogen concentrations and suggested most symptoms of perimenopause derived from it. People in their forties can start to experience increasingly irregular cycle lengths, heavier than usual periods, and vasomotor symptoms like night sweats or hot flashes (lactating can also cause some of these symptoms, which can be compounding for people like me who breastfed into their early forties). However, this is not occurring from low estrogen but more likely bouts of higher estrogen (remember, as we age our ovulation can get a bit extra) and sometimes also lower progesterone (from the also frequent bouts of anovulation, or lower-quality corpus lutea from older follicles).[24] As someone just starting to go through perimenopause, I wish violence upon my estrogen-to-progesterone ratio.

Another term some of us hear from our gynecologists, issued in warning tones, is the status of our *ovarian reserve*. There are several imperfect ways to assess one's ovarian reserve, most commonly measured via FSH or anti-Mullerian hormone (AMH), but you might sometimes also get an antral follicle count via ultrasound. These are all deeply imperfect estimates of how many eggs one has left and what quality they may be.

AMH is the most common ovarian reserve biomarker because it is produced by granulosa cells in the earliest stages of follicle development; a biomarker for these cells therefore represents whatever undeveloped follicles we may have to recruit for ovulation. As one approaches menopause, the culling of early follicles ramps up significantly, perhaps because so many of these follicles that are left are not in the best shape. Some of the growth and differentiation happening even at these stages is compromised when these follicles start getting old. For instance, the risk for aneuploidy (when you have an extra chromosome or you are short one) goes up with age. This increased risk comes from a deterioration of the processes of chromosome segregation and a shift in the main location of meiotic crossovers toward the center.[25]

Menopause constitutes that moment when the ovarian reserve is more or less gone, and processes in the ovary are pretty dysregulated such that no more menstrual cycles can occur. There can be several years of fits and starts that precede that full year of amenorrhea required for the clinical definition, when one can go months between periods and almost start to think the whole thing is finally over. Few of these cycles are ovulatory, and at this later stage of perimenopause, there probably is fairly low estrogen. Across most industrialized populations, menopause occurs pretty reliably between fifty and fifty-one years of age; among less industrialized populations, around forty-eight years of age. In neither case is there the wide range of variation that we tend to see in menarche, which stretches from nine to seventeen years.[26]

While age at menopause does not have as wide a range as menarche, there are some factors that appear to influence its timing. The number of children plays a role because all that time being pregnant and lactating is time you are not ovulating more eggs—but only to a point. In one study of over three hundred thousand postmenopausal people in Norway, researchers found a positive correlation between menopausal age and number of children, but only up to three kids.[27] Beyond three kids, there was no further delay of menopause. There are a few other variables that affect menopausal timing, like smoking and physical activity.[28] These factors can increase or decrease the number of available oocytes to ovulate and the rapidity of ovarian aging. The fact that we make so very many eggs at once, and that we cull them in so many different ways before and during our potentially reproductive years, shows just how discerning our ovaries are in deciding which eggs to select.

If, as I have suggested, the ovaries are brilliantly bossy and continue to be so through their entire lives, it stands to reason that the eggs must be bossy, too, as the next major bottleneck for reproduction occurs when an egg joins up with a sperm. The uterus is also involved in this decision, seeing as the roommate absolutely has a stake in who else might be moving in. When you are the one housing an embryo and supporting its growth, development, and eventual independent life—when your body is the one feeding that body not only through gestation but often through lactation for months and years afterward—it makes sense to have evolved multiple checks on whether one should put in all this effort.

THE HEROINE'S JOURNEY

If you went to middle or high school in the United States, you likely remember Punnett squares, the diagrams used to chart dominant and recessive traits in things like Mendel's peas in order to predict what percentage of a crop will display certain characteristics (yellow versus green, wrinkled versus smooth shells). The tidy percentages we get from those calculations—25 percent of one, 50 percent of another—have never perfectly matched up with the real numbers in a given population.

Both experiments and observations of gene pools have revealed what are called transmission ratio distortions, where we see deviance from the expected ratio of different genetic variants in a given population. At first, researchers thought that the final ratios in the populations they observed were different than expected because some of the variants couldn't survive and therefore were being miscarried. However, mouse research has shown that we don't see miscarried embryos or fewer total offspring born to mothers in many cases of these "distorted" genetic ratios. Why, then, are those Punnett square percentages mistaken?

As it turns out, eggs and sperm have opinions about their choice of partner, and the process of gamete fusion is less random than we once believed.[29] However, the fact that bodies with uteruses are the ones with the greatest ability to control whether and when to reproduce goes against our strongest cultural dictates of what constitutes a normal (cisgender, white, straight) female: passive, subservient, willing.* The anthropologist Dr. Emily Martin points out the allure of this narrative in her aptly titled article "The Egg and the Sperm: How Science Has Constructed a Romance Based on Stereotypical Male-Female Roles." The ovary, having already produced all the eggs it will ever need before birth, is viewed as passive—just waiting for her prince. The testes are a site of production and activity, a capitalist's dream: they make sperm at a constant, dizzying rate.

These hero narratives persist not only in the journal articles and book chapters we scientists write for each other. According to an analysis of seventeen major medical anatomy textbooks published since 2008, women are still disproportionately described performing passive activities.[30] If medical students are instructed with textbooks that promote gender stereotypes and diminish the experiences of women—not to mention render invisible the experience of genderqueer and nonbinary people—medical researchers are in danger of bringing these stereotypes into their formation of hypotheses, selection of methods, and

* As I have pointed out before, selectivity in a biological process is not the same thing as individual choice: I want to be clear that the body's inhibiting or promoting processes that can lead to pregnancy has nothing to do with what that person may want. As always, these types of bodily and personal autonomy operate together.

interpretation of their results. Perhaps even worse, medical doctors trained by these books risk bringing these stereotypes into their clinics and hospitals.

When I tried to read up on some of the interactions that lead to gamete fusion, I was struck yet again by the way that the sperm is still situated as the explorer and actor. In one review cited by over a hundred other papers, the authors describe the quest of the sperm "to find a single cell—the oocyte. The spermatozoa subsequently ignore the thousands of cells they make contact with . . . until they reach the surface of the oocyte. At this point, they bind tenaciously to the acellular coat. . . . These exquisitely cell- and species-specific recognition events are among the most strategically important cellular interactions in biology."[31] The ability of sperm to differentiate between an egg and the other cells in the vaginal tract and uterus is a notable fact. But, I don't know, unless one is already in awe of the noble sperm, it's not *that* cool? A wide variety of cells in our body engage in *cell- and species-specific recognition*. As I write this paragraph, my stomach is currently distinguishing between the contents of a Greek salad with chicken and several strains of bacteria and its own lining, a pretty astounding array of different species to recognize and treat as food source, friend, foe, and self. I'm not sure I'd say my stomach is full of exquisite and strategically important events. It's just digesting my food.

As we will now see, the process of reproduction is much more interesting and complex than the simple narrative of feminine passivity would lead one to believe. Instead, we see a heroine's journey less about overcoming obstacles and more about cooperating with others. First, the uterus prepares for the possibility of pregnancy. During the follicular phase, the endometrium thickens, undergoing substantial remodeling of the glands, vessels, and connective tissue. This phase of endometrial development is called the proliferative phase because of the ways in which most of these processes serve to increase the mass of endometrial tissue. After ovulation, when the corpus luteum is more in charge, the endometrium starts to differentiate into the sort of place an embryo might want to hang out for a while. Subnuclear vacuoles—similar in some ways to follicles, as they are fluid-filled sacs—line up along the glands of the endometrium. Their job will be to enhance the secretory activities of the endometrium. Arteries are coiled: dubbed *spiral arteries*, they maximize oxygen and nutrient transport. The connective tissue of the whole thing—the endometrial stroma—becomes denser, in part to better control the process of interfacing with a possible embryo.

The uterus builds up this quality environment and creates opportunities for discernment so that should conditions be right and a worthy sperm be chosen, the timing of these events will cascade appropriately in order to give the reproductive process the greatest chance of success. For instance, we know that embryo-uterine interfacings occur six to twelve days after ovulation, but the vast majority—over 80 percent according to one study—occur in days eight to ten.[32] This tight window is made possible by a number of processes and backup processes that can, if necessary, rearrange the body's schedule.

First, there are the aforementioned follicular waves, which enlarge the window for appropriately timing ovulation. This is the reason that the follicular phase (the time from menses to ovulation) is so much more variable in length than the luteal phase (the time from ovulation to the next menses). Next are the cervical crypts, spaces where the uterus seems to shunt and store sperm to use later with nice, sticky cervical mucus in order to prevent overcrowding at the egg and allow for some selection of preferred sperm. In the only study to examine this process in humans, published in 1980, women consented to be artificially inseminated the day before their scheduled hysterectomies. Researchers then cut each cervix from the removed uteruses into three parts—the upper, middle, and lower cervix—which they sliced into sections six micrometers wide (one micrometer is one ten-thousandth of a centimeter). They microscopically examined every tenth of these sections for endocervical crypts and the presence of sperm. The researchers found that the location and size of cervical crypts varied according to whether the participant was pretreated before insemination with estrogen or progesterone; they also found that when the participant was inseminated with abnormal sperm, not much was stored in the crypts, and the little that was stored appeared only in the lower crypts. There are still a number of unanswered questions about crypt storage and release mechanisms, but many reproductive tracts across many animals have ways to store and save components of reproduction for later use.

In addition to the follicular waves and cervical crypts, there are also uterine waves, a special type of muscle contraction that helps control the speed at which sperm reach the egg, propelling them on a journey that would otherwise be too long for them to make on their own.[33] Early in the cycle, these contractions move from fundus to cervix, discouraging sperm or other invaders from making their way up the reproductive tract. Later, though, when gamete fusion and interfacing may be desirable, uterine waves switch direction, from cervix to fundus, so upward and inward. The final stage of sperm maturation happens in the vaginal tract and uterus—it is only here that sperm gain their full ability to move about, and the egg's outer coating, the zona pellucida, exerts an awful lot of control over whether and when a sperm can join with it. Four different zona pellucida glycoproteins contribute to this process, and once the one sperm has been chosen, a near-instant depolarization of the membrane occurs to prevent any other sperm from trying to join up.[34]

We're still not done: the uterus provides additional layers of control over timing and selection. Better sperm tend to "stick" to the fallopian tubes to develop a sperm reservoir.[35] And even after the egg selects a sperm, the endometrium offers up one more layer of embryo selection. If a worthy embryo has been formed, the endometrium guides it to the exact point where it would like to begin interfacing; if the embryo is inadequate, the endometrium disengages its receptive processes. Finally, through differentiation of its tissues, the endometrium creates a dense matrix that provides some feedback for the trophoblast (the next stage of development for the embryo) as it moves in, protecting the parental body and its interests in future reproduction by not letting the trophoblast get too deep into the endometrium so as not to risk giving the fetus too much control when it comes to resource allocation.

In addition to physical barriers in the tissue itself, the endocannabinoid signaling pathway creates another layer of choice and control for the uterus. While the term "cannabinoid" refers to substances we often find in cannabis, or marijuana, it turns out that we make these substances too ("endo" is a prefix to imply something produced within the body). In the reproductive tract, endocannabinoids feature at a few key times: they appear to be related to uterine contractility, transport of the embryo to the endometrium, and signaling between the endometrium and embryo to determine the exact place to implant. They are also part of the final detection system for ill-formed embryos, where they influence the receptivity of the endometrium to prevent embryo-uterine interfacing.[36]

If for whatever reason no suitable sperm are selected or there were none around that cycle, the corpus luteum in the ovary has no reason to stay, as its main job is to supply progesterone to help orchestrate early pregnancy. Progesterone concentrations decline, and soon we've got blood in our underwear. Once we appreciate how many different processes take place to ensure proper selection, the idea that a normal menstrual cycle always looks a certain way—a fourteen-day follicular phase, ovulation, a fourteen-day luteal phase—starts to seem less realistic. A person might end up with a menstrual cycle with no ovulation, multiple ovulations, a longer follicular phase, or a very short luteal phase, and this does not immediately mean anything is wrong! For this reason, I want to conclude with a short story about a dorky collaboration that is upending our idea of what a normal menstrual cycle should look like.

NORMA'S MENSTRUAL CYCLE

In my lab, we see a lot of menstrual cycles. We have conducted menstrual cycle research at a summer science camp in the United States to understand adolescent cycles and on working scientists to understand how cycles are affected by the stressors of the competitive and often hostile science climate. Our largest samples come from a comparative study between rural Polish and Polish American women. Even these menstrual cycles—constrained by age, lifestyle, and sometimes even ancestry—are hugely variable. In order to conduct statistical analyses on the hormones we measure from blood, spit, and pee, we usually need to use some kind of formal method to align all the different individual menstrual cycles. Most of the time, we use what is called the midcycle drop date method. This method, adapted from that of my undergraduate mentors Drs. Susan Lipson and Peter Ellison, starts by finding the *drop day*, defined as the second of two consecutive days around midcycle during which there is the greatest decrease in estradiol. While lab alum Dr. Katharine Lee, now a professor of anthropology at Tulane University, wrote a computer program to identify the drop dates of our recorded menstrual cycles, there were still a large portion of cycles that needed to be coded by hand.

Katie, Mary, Meredith Wilson, and I spent multiple meetings going over these menstrual cycles, trying to decide where the drop day was. Based on the presence of progesterone, we knew that we were only looking at ovulatory cycles (anovulatory cycles typically have very low progesterone since there is no corpus luteum to make it). Yet we saw ovulatory cycles with double peaks of estrogen at midcycle. We also found cycles where the luteal-phase estrogen was higher than the midcycle estrogen (traditionally, midcycle is where you see that "peak" before the "drop"). For a significant subset of our research participants, traditional expectations about how a menstrual cycle is supposed to work were rather unhelpful in locating their midcycle.

Throughout the broad scope of our research, we have never found any Normas—people whose hormone values look just like the textbooks. Indeed, we have not even found many Martha Skidmores. We can replicate the "normal menstrual cycle" with our data by averaging everyone's values, just like those statues, but no individual's hormones match those averages.

We are hardly the first lab to notice that the menstrual cycle varies. But the idea of normality has become so powerful, so entrenched, that much biological research involves averaging toward a mean phenomenon in such a way that it may not be describing anything that is actually real. Instead of there being one norm, from which everyone deviates, what if there were a number of normals? In other words, what if there were multiple menstrual cycle phenotypes—different hormone patterns— that related to demographic differences (like age and parity) or lifestyle differences (like physical activity)? To some extent, we already know that this is the case because, in our research, we can see that eighteen-to-twenty-four-year-olds, say, tend to have lower hormone concentrations

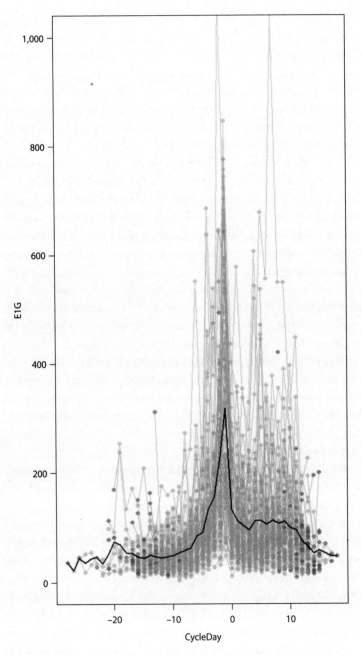

The estrone-3-glucuronide (estrogen metabolites) from ninety-eight Polish and Polish American ovulatory menstrual cycles. Averaging all values looks like what you might remember from a textbook, but check out all that variation.

than twenty-five-to-thirty-four-year-olds.[37] But even here, we tend to pay attention to amplitude, or total volume of a hormone, rather than the shape of that pattern over time.

One day, when I had been pontificating (read: complaining) about this to my students, Merri admitted that she had been talking to her college friend and fellow anthropologist Dan Ehrlich about this idea for a while and that they thought there might be a way to group menstrual cycles by pattern using geometric morphometrics. Put simply, geometric morphometrics studies the shape of things—the shape of a mouse's hip bone, or a fish, or a hominin fossil. You set a certain number of landmarks on the thing whose shape you want to understand and then you can make shape comparisons and sort samples by overall morphology. You can even scale the measures to raise the importance of shape over size so that, for instance, specimens that are from the same species but different sizes will be grouped together, while similarly sized but very differently shaped objects will not be. Though the smallest jammer on my roller derby team is probably five feet tall, while our largest blocker is closer to six feet, geometric morphometrics would be able to identify all of us as human by the shape of our hip bones.

Geometric morphometrics allowed us to study the pattern of hormones without overfocusing on amplitude—a real problem when studying estrogens, in particular, because the big peak that occurs midcycle often statistically wipes out variation everywhere else. Dan started coming to our Zoom lab meetings, another lab group member Valerie Sgheiza came up with a clever name, and the moRphomenses collaboration was born (the capital R refers to the code Dan and many others in my lab use to run statistical analyses). Using our data set of Polish and Polish American women as our test case, we found that our sample contained three distinct phenotypes of estrogen patterning through the menstrual cycle and as many as ten subtypes. We have a skinny midcycle peak, high luteal estrogen group (Group A); a wide peak, high luteal estrogen group (Group B); and a skinny peak, moderate luteal estrogen group (Group C). Group A looks the closest to Norma's estrogen pattern, yet it has the smallest number of people in it, where Group B is the largest.

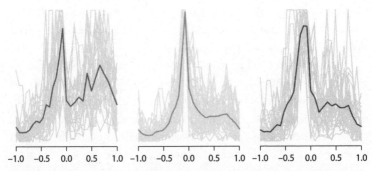

The three estrogen patterns from the ninety-eight cycles in the prior figure.

The central idea here is not that there was one Norma and now there are three; it's that there are a wide variety of definable patterns that are likely to be sample and even population-specific. Until recently, averages and norms, even population-specific ones, kept us from conducting more sophisticated analyses on the pattern of hormones through the menstrual cycle. They kept us from noticing the strong-willed leadership of the ovary, the tactical protections of the uterus. They have even led people well within that "normal range" of age at menarche to self-define as "early" or "late" bloomers.

In the next three chapters, we will continue to contradict both the idea that there is a mean that represents the best or most healthy, the pervasive myth that those things we gender feminine are passive, and its equally pervasive corollary that these mechanisms should be gendered at all. To do so, we will turn our attention to the subject of environmental stressors. I will describe the ways in which the stressors we face—from food and exercise, from immune challenges, and from psychological stress—influence the menstrual cycle. Some would like us to believe that having a menstrual cycle makes us hysterical and sensitive, and others may want to gaslight us into believing our experiences of our cycles are not meaningful. The scientific reality is far more nuanced.

CHAPTER 3

The Stress of Energy

IN 1873 in his book *Sex in Education; or, A Fair Chance for the Girls*, Dr. Edward H. Clarke put forward the idea that women should not be encouraged to seek an education because it would affect their fertility.[1] The demands of higher education, he argued, would deprive their reproductive organs of energy, causing "their ovaries and uteri to shrink," and he based this theory on a measly seven case studies, at least one of which was demonstrably false.[2] Clarke was yet another eugenicist, worried about the decline in the birth rate of white people and the masculinization of white women into a "sexless class of termites." He was also vexed about the possibility of coeducation at his precious Harvard.*

As the historian of science Dr. Naomi Oreskes has written in her book *Why Trust Science?*, while ideas about the delicacy of women, in particular white upper-class women, were not new at this time, Clarke's *limited energy theory* provided a scientific narrative that many were more than happy to use to argue against the many women's colleges that were cropping up, as well as the movement for white women's suffrage.[3] What often gets lost is that in fact Clarke had many detractors at the time, including Lucy Stone, Julia Ward Howe, and Dr. Mary Putnam Jacobi. They argued that if girls were more prone to illness, it was because they were denied outdoor play and given additional domestic and social responsibilities compared to boys. The educator Anna Callender Brackett

* Sorry my dude, class of 2001!

declared that "it is the lounging, deadening brain work of which we have too much, not the active, vivifying brain work of which we have too little that does injure the system." The painter and feminist activist Eliza Bisbee Duffy remarked that Clarke "knows if he succeeds in carrying the points which he attempts and convinces the world that woman is a 'sexual' creature alone, subject to and ruled by 'periodic tides,' the battle is won by those who oppose the advancement of women."

Despite the efforts of these and many other opponents, ideas akin to the limited energy theory continue to influence our understanding of the physical capabilities of people with uteruses, as well as our understanding of the effects of physical activity on ovarian and uterine function. Clarke's hypothesis derived from the first law of thermodynamics, which states that while energy can be transformed from one form to another, it can neither be created nor destroyed. Curiously, his idea that the body is a closed energy system—where energy used for the brain cannot then be used for reproductive organs—was primarily applied to women.

One popular way to illustrate the laws of thermodynamics is with a story about a child's misplaced blocks.[4] In this story, the child loses twenty-eight blocks, and the mother recovers twenty-seven. Even though the mother can't find the missing block, she knows that it must be somewhere in the house. Similarly, even if we lack the tools to measure energy transfer in a way that confirms the laws of thermodynamics, we know that in a closed energy system, the missing energy must be somewhere. With respect to the human body, if someone consumes more calories than they expend, we expect to see that energy appear as increased reproductive hormone concentrations and possibly also increased body mass. If someone consumes fewer calories than they expend, we expect that they will show lower reproductive hormone concentrations and perhaps lose body mass.

We used to believe that if we measured energy expenditure and energy intake and the calculations did not turn out right—that hormones or weight or fat percentage did not change in the expected way—then we had made an error in measurement. We were like the mother searching for the twenty-eighth block. This was not an entirely unreasonable assumption: people have a tendency to underreport how much they eat and, in some conditions, overreport how much they move around.[5] However, even as scientists have adopted advanced technologies to overcome concerns about reporting biases, they are still not always able to find the twenty-eighth block; that is, they cannot reconcile our common understanding of movement and nutrition with the kind of effects we see in a person's body.

In later chapters, we will complicate the idea that energy in and energy out have a straightforward set of effects on the body. However, in this chapter we will work within a simplified model in order to begin to understand how energetics affects the menstrual cycle. This model assumes the body has three main needs: to support maintenance, reproduction, and growth. Early in the lives of humans, our needs are two: those of maintenance, which includes the pulmonary, cardiac, neurological, immunological, and digestive efforts that keep us alive, and growth, which includes skeletal and organ growth and the maturation of various systems.

Around puberty is when our third need emerges, and for a period of several years, human bodies must figure out how to provide enough resources to meet maintenance, reproduction, and growth needs. Reproduction is concerned with the maturation of the reproductive system, including both immediate issues of fecundity (the ability to become pregnant) and longer-term considerations of whether one has enough energy to support reproduction for many months of gestation and lactation. Once we have finished growing, the growth need goes defunct, and we only concern ourselves with allocation to maintenance and reproduction. Should there be reproductive senescence, like menopause, or removal or hormonal neutralization of the gonads, then we are down to just supporting maintenance needs.

Because this is not a closed system, the question becomes: What happens to reproduction when there are constraints on our access to resources from high-energy expenditure, low-energy intake, or both? What kind of variation do we see when we are unable to always fully support that reproduction need? And what happens when some individuals have a surplus of energy beyond their needs? Throughout our discussion of the effects of variation in energetic resources on the menstrual cycle, we will continue our interrogation of the Western understanding of normative femininity—especially the concepts of femininity as tidy, passive, and caregiving. These concepts, which have served to limit the movements of women and femme people, have led to two major misunderstandings of the basic ways that energetic resources are allocated.

First, the notion that women are fragile has led to long-standing worries that intense exercise and significant energy expenditure are harmful, which has obscured the much greater risk that societal body standards lead many people but especially women and femmes to undereat.[6] Second, when we conceive of femininity as giving, we see reproduction and the care of others as women's highest priority. In doing so, we make incorrect assumptions about how energy is used throughout the reproductive process. Ovarian function can and does anticipate quite a bit of energy expenditure without any lasting or harmful effects, and it protects and prioritizes itself over any other beings for which it might be responsible. As a result, the body frequently deprioritizes the menstrual cycle in a way that is flexible and adaptive, that leads to a range of experiences, and that allows it to spring back as conditions and desires allow. Put another way, menstrual cycle variation stemming from energetic variation is a feature, not a bug.

ALLOCATING ENERGY

There is an organizing framework to this prioritization process of how to allocate energy to maintenance, reproduction, and growth. Life history theory is the concept that all organisms, faced with limited resources, have to "decide" how to allocate them. At different stages in our lives, with different commitments or values or dependents, we confront different trade-offs around how to allocate our time and resources. The intent is to maximize reproduction in the long term, to have as many babies as possible who survive to adulthood and make their own babies. Between different species, and even between different individuals within a given species, the optimal allocation will vary considerably. It will depend, for instance, on phylogeny, meaning the different traits available based on a species' ancestry; for humans, this involves the fact that we are big-brained, bipedal, and have long childhoods and long lives, as well as some interesting specifics related to how our uteruses and placentas work. It also depends on an individual's access to resources; for humans, this can refer to time, money, food, shelter, education, friendship, physical and emotional safety, and much else.

Life history theory helps us understand big-picture stuff, like why we might perform unselfish acts, such as watch for predators on behalf of the group or care for children who are not our own, that are baked into many human cultures. We've invested in big brains, which require an extended childhood to grow those brains and use them to learn. This explains, in part, why we tend to have one kid per pregnancy. But even raising one needy booger at a time (please tell me I'm not the only one who calls my kids that) is too much for any one or two parents to handle, so we have evolved cooperative breeding practices that involve the support of friends, grandparents, siblings, and others.[7] Labor support, postpartum support, and shared childcare are crucial to raise human kids, which is why for modern parents in nuclear families, living away from family can be incredibly difficult unless they can afford to pay for childcare and other forms of support. Parents who do not work outside the home, even parents who fully homeschool, still often make use of modern infrastructure that replicates cooperative breeding: they use online classes, museums, zoos, parks, and libraries (and the staff who work there) to augment their labor.

Life history theory also helps us understand smaller stuff, like why we might not ovulate in a given menstrual cycle. As we have already discussed, Norma's menstrual cycle—an ovulatory, twenty-eight-day cycle with one acceptable hormonal pattern—does not exist for any specific individual. To take it a step further, the deeper problem is the idea that Norma's menstrual cycle serves as an ideal menstrual cycle: doing so presumes that it's always a good idea to devote significant time and resources to the menstrual cycle. In other words, it presumes that a normal body would preferentially allocate resources to creating fecund menstrual cycles at the expense of other processes in the body. This idea—that reproduction and thus the menstrual cycle are the body's highest priorities—is part of the foundation on which Clarke's limited energy theory was based.

Our bodies understand that most of the time, prioritizing the menstrual cycle and thus our fecundity is not a great idea. From the perspective of the one with the uterus, reproduction involves nine months of gestation and, for many, several years of lactation. It also involves the intensive demands of parenting described above. Taking these costs into account, the idea that one should be constantly at the ready for pregnancy makes little sense. To maximize your Darwinian fitness, you only want your body to signal availability for reproduction when you possess enough resource to handle its costs—not merely the energy needed for a menstrual cycle but the energy needed for years and years of pregnancy, lactation, and parenting dependent offspring.

While our overall Darwinian goal might be to maximize reproduction, depending on your age, rank, sex, or other factors, the best way to pursue that goal may be to delay or avoid childbearing. Whether or not the body is "thinking" about any of these things, it nonetheless must make trade-offs between different goals. Our bodies might consider the trade-offs between reproducing and surviving, between the children we already have and the possibility of future children, or between having a bunch of kids or investing more in just one or two. Thinking about energy as context dependent and finite helps us understand that, even in a system where reproduction appears to be the reigning queen, we might not tithe all our earnings to her. And when there is any amount of constraint on the amount of energy entering the system, that tithe is the first thing we will short.

ENERGY IN, ENERGY OUT

How much energy you have available to put in the maintenance, growth, or reproduction barrels is dependent on two basic factors: how much you eat and how much you move around. This is, at least, a good first way to think about energetics. The energy immediately available to us comes from the food we have just eaten in the form of glucose; we also have access to the energy we store in muscles via glycogen and that we place in even longer-term storage mechanisms like fat. To understand what amount of energy someone might have available to them, we might look at their food intake and physical activity patterns, their glucose or glycogen, or their fat storage and distribution. To obtain this information, we may use surveys, observations, and blood tests. For direct and quantitative measures, we have dual-energy X-ray absorptiometry scans and BodPods (machines that measure your body composition). You may even have a device at home, a bioimpedance scale, that you can stand on to get an estimate of your percent body fat based on some assumptions about your age, gender, activity level, weight, and height.

This way of thinking about energy—how much energy can you recruit when needed—is known as energy status. It's a snapshot of what a person has available to them that moment. It's also the one clinicians seem to rely on the most when making subjective decisions about the health of their patients, such as when they plug our weights and heights into a body mass index (BMI) calculator (which, while you're here, I might as well tell you was originally conceived in the nineteenth century by Adolphe Quetelet, a physicist whose concept of the "normal man" strongly influenced both anthropology and eugenics). The second way of thinking about energetics is energy flux, which is the rate at which energy is moving through you. An example of a high-energy-flux person is an Olympic swimmer like Katie Ledecky, consuming thousands upon thousands of calories a day but also expending as many with punishing workouts. A low-energy-flux person, by contrast, is someone who is very sedentary but also consuming little, like my grandmother during her final years. You can imagine that Olympian Katie Ledecky and my grandmother may have had similar amounts of body fat, but that understanding of their available energy only captures part of the story.

The third way to think about energetics is in terms of energy balance, which is based on whether you have eaten more calories (positive balance) or expended more calories (negative balance) over a given period of time. Whereas energy status is like a screenshot, energy balance is like a GIF, as it gives you a sense of change over time. If you are logging your food intake and physical activity each day, you might get a sense of whether you are in energy balance for that day. If you are measuring someone's body fat composition, you will capture their change in energy balance over a longer period.

As the study of metabolism and nutritional science is increasingly coming to acknowledge, this basic understanding of energetics—that we can get a measure of a person by measuring calories in and calories out—is incomplete. The number of calories we can access in food can depend on whether we cook it, the timing of what we eat matters to how it is processed in the body, and the metabolic effects of different activities can vary independent of how many calories they lead us to expend. While we once thought metabolism was a matter of simple algebra, it's really more like multivariate calculus. Emerging evidence suggests, for instance, that the physiological effects of sitting around on our butts are not necessarily benign or neutral, even if we tack on thirty minutes of exercise at the end of the day.[8]

This simplistic way of thinking about energetics—as calories in and out—has also supported the mistaken belief that greater energy storage must mean too little movement or too much eating. Fat is not all bad nor is the relationship between fat storage and energy as straightforward as most people believe. It is easy to store significant amounts of fat while leading an active lifestyle, and substantial research supports a stronger relationship between physical activity and health than between body composition and health.[9] Our bodies very much want to retain fat—a safeguard against drought or famine—and fat is critical to reproductive function, as well as to parental, fetal, and infant health. Later in this chapter, I will describe how our misguided belief (largely from fat stigma, which, you guessed it, also stems from eugenics!) that fat storage is itself a health problem has obscured our ability to understand certain illnesses related to the menstrual cycle.

To recap, a calorie is not always a calorie, fat is not always bad, and we need to distinguish between energy status, energy balance, and energy flux if we want to understand how we support our three possible bodily needs of maintenance, reproduction, and growth. For instance, high energy flux—someone eating food like a Kardashian makes money and expending effort like a Kardashian spends money—can create a greater sense of uncertainty about the availability of resources, which can constrain the availability of reliable energy for that reproduction bucket. That said, energy deficits are the biggest determination of whether you can fill your reproduction barrel and therefore can have the greatest effects on the menstrual cycle.

CRITICAL THRESHOLDS

To understand how our bodies decide to allocate their energy budget, I want to go back to a particularly tricky time for humans: puberty and adolescence. As I mentioned earlier, puberty is the time in one's life when resources are needed for maintenance, growth, and reproduction at once. Humans can continue to grow for several years after puberty, marking a period of substantial overlap in their bodies' need to get larger and get the reproductive system up and running. By measuring energy, growth, and the timing of certain reproductive system benchmarks, we can learn how or why a body might choose to allocate more resources to growth than to reproduction, or the other way around.

In my field, the most famous example of this research was conducted by Dr. Rose Frisch, a scientist who in the 1970s studied the effects of body fat on the timing of menarche and, later, fertility. Frisch died in 2015, and in her Harvard School of Public Health obituary, her son describes the substantial sexism she faced in her career.[10] Frisch held a PhD in genetics and had even worked as a "computer" for Richard Feynman, but she left science for many years to raise children. When her youngest was in seventh grade and Frisch decided to reenter the workforce, she was at first only able to secure a research associate position, rather than a faculty job, at the Harvard Center for Population and Development Studies. Her race and class surely helped her maintain the toehold she did have: married to an MIT professor and thus knowing they had a stable income, she took the kind of position that did not pay the bills and followed her passion. As her son Henry tells it, even once she was a professor her salary was abominable: once "she got a phone call from the National Institutes of Health that her salary on a grant application was supposed to be her annual salary and not her monthly salary, to which she replied, 'That is my annual salary.'"

The bias against Frisch did not manifest only in her difficulty securing a job or receiving fair pay. Male scientists had very strong reactions to Frisch and her work. They accosted her at conferences and wrote public criticisms of her work that often attacked her as a person, rather than the hypotheses she and her collaborators put forward.[11] The abstract of one paper, with the title "Statistical Flaws in Evidence for the Frisch Hypothesis," dismissed her work as useless:

> Frisch has drawn attention to the role of fatness as a determinant of the onset and maintenance of menses. Her index of fatness, upon which her conclusions rest, contains such a large component of error that it cannot serve as an accurate measure of fatness. Hence, claims about the role of fatness cannot be supported. Specifically, the procedure for estimating age at menarche, which is based on first classifying girls into quartiles of fatness, can be shown to be useless.[12]

The cumulative degree to which this work and others derisively attacked a colleague is unlike what I see when scientists engage with ideas that originate with men.

What idea did Frisch put forward that was so noxious? Frisch hypothesized that, in order for young people to achieve menarche, they needed to reach a critical threshold of body fat—an idea she first based on the fact that very malnourished girls tend to hit menarche when they are far older than average. The timing of menarche is thus predicated on a young person's energy availability and energy status. To test the hypothesis, Frisch used population-based data to estimate fat reserves alongside girls' ages at menarche. With the measures she used, there were some statistically significant relationships: she found that a minimum weight was necessary to achieve menarche and that a threshold weight was necessary to resume or maintain menstrual cycles in girls who had stopped cycling.[13]

Many scholars, using more direct measures of fatness than Frisch did, have found that Frisch's hypothesis does not hold up to further scrutiny. Further, a broader examination of variation in girls—rather than a very narrow definition of normal adolescent girls—has shown that this critical threshold predicts too late an age at menarche in taller girls and too early an age at menarche in fatter girls. More generally, the fact that we are all so variable—recall, Norma exists in museum statues and medical textbooks but not in real life—undermines the likelihood of such a thing as a fatness threshold. That said, while Frisch's hypothesis was not supported, it was a buildable idea: science is a process rather than a static set of facts, and Frisch's work helped open the door to thinking about pubertal timing and, eventually, menstrual cycles as variables that could be influenced by energetics.

Frisch's work raised, for instance, the more general question of whether there are any critical aspects of the body that must be grown before transitioning from a focus on growth to reproduction. One person who took up this question was my undergraduate adviser, the anthropologist Dr. Peter Ellison, who compared the critical fatness threshold hypothesis to another physical measure: skeletal maturation. Ellison contended that, if any critical threshold existed necessary to initiate menarche, it was probably related to the size of the pelvic canal so that a baby would be able to pass through it. According to this hypothesis, overall size (relative to population) and hip width (as a proxy for the pelvic canal) should correlate with menarcheal timing. In his analysis, weight was still important, just as with Frisch's initial work. But skeletal growth, especially hip width, was a far better predictor of menarcheal age.[14] (Ellison has a lovely and far more thorough description of this work in his book *On Fertile Ground: A Natural History of Reproduction.*[15])

Given the plausibility and importance of Frisch's work, why were contemporary scientists so hard on her ideas? There are two interrelated reasons. The first is that her hypothesis challenged feminine norms. By arguing that fatness was necessary before menarche, Frisch suggested that bodies consider their own needs in the timing of reproductive events and of course implicitly equated fatness with femininity, a major no-no when our cultural values by the twentieth century encouraged thinness in women. This suggestion stands in stark contrast to the narrative of normative femininity, according to which women ought to prioritize giving to others (often at a personal cost).

The second reason was that Frisch was a woman, interceding in what had previously been a largely male space. The sciences were (and remain) male dominated, and the theoretical sciences where people put forward their Big Brilliant Ideas were (and remain) especially so. As Oreskes has written, science as it is understood today is a process of correction.[16] That is, interrogation, integration, peer review, and other forms of criticism and engagement are what drive the scientific process. The process of science, the fact that it is eminently buildable and changeable, is what I love about it. However, scientists come with their own lived experiences and biases; without sufficient diversity of thought, those powers of criticism can oversample one point of view. In both the study of human evolution and medicine, when those who are underrepresented in those fields dare to present big ideas, to make their own corrections, they often find themselves criticized more harshly and remembered less fondly than their white male peers. We teach Lamarck and Mendel to middle school biology students, not because their ideas are still fully supported but because their hypotheses help us understand how we eventually arrived at Darwin and DNA. Frisch and Profet—who you may recall came up with the sperm-borne pathogens hypothesis and upended the way we think about the evolution of menstruation—should be recognized in the same way. While Frisch was mistaken in thinking that there is a critical threshold for fatness for menarche, her research led us to understand that fatness is important to the menstrual cycle more broadly. The energy we are able to store is a signal to our bodies that we have sufficient energy for the long and costly process of reproduction—from ovulation and implantation through gestation and lactation. How much fat is required is context-specific, rather than absolute: a lifetime athlete's signal that she is able to store energy is different than a forager's, which is different than a farmer's, which is different than a video gamer's.[17] So Frisch wasn't wrong; she just wasn't right.

PHYSICAL ACTIVITY IS DEAD, LONG LIVE PHYSICAL ACTIVITY

While Frisch's research drew important attention to the idea that energetic constraint—low energy status or negative energy balance—causes variation in fecundity and fertility, it also, unfortunately, contributed to the limited energy theory–heavy clinical perspective that (white middle-class) women were constantly at risk of harming their fertility by engaging in any behaviors that might overexert themselves and reduce their energy available for reproduction. Over the years, many have resisted these assumptions and engaged with all manner of outdoor sport, ranging from long walks to tennis to more vigorous team sports. The reactions some have to the moments when women may turn out to be competitive with men belie the truth of these claims of protection of fertility. From Clarke's screed intended to prevent women from going to Harvard, to bans on women runners using men as pacers to set world records, to trying to narrow the definition of women to exclude trans women from competition, the goals in fact are protection of the status of men and control of the living conditions of women and gender minorities.[18]

My colleague Dr. Jerrilyn Prior has something to say about this. Dr. Prior is a Canadian health researcher and the director of the Centre for Menstrual Cycle and Ovulation Research. Prior herself was inspired to study menstrual health and physical activity back in 1967, the year Kathrine Switzer became the first woman to run the Boston Marathon as a numbered entrant (bandit racer Roberta Gibb had beaten her time by over an hour the previous year). The safety of women running marathons was hotly debated at the time—the claim was that the female constitution was too weak for endurance running of that distance—and Switzer's audacity to challenge this belief provoked significant backlash. The Boston Athletic Association director at the time said of Switzer running, "If that girl were my daughter, I would spank her."[19] During the race, Switzer was physically assaulted and had to be protected by her boyfriend, who ran alongside her. As these responses demonstrate, many men were less concerned with the safety of women than they were in maintaining the sport as a male-only space and dictating what constitutes appropriate behavior from women.

Five years after Switzer ran, women were officially allowed to enter the Boston marathon, though women were not allowed to run the marathon in the Olympics until the 1984 games. Even today, many sports, like ice hockey and lacrosse, have different rules for women and men, with most of these rules limiting the movements of women. Women's and

Image of Katherine Switzer being physically assaulted by
Boston Athletic Association Coach Jock Semple while she is
running the 1967 Boston Marathon.

men's swimming events only equalized for the 2020 Olympics, before
which the assumption had been that women could not safely compete
at the fifteen-hundred-meter distance.

Observing the reaction to the Boston Marathon and similar events,
Prior perceived that the medical community had a biased perspective
about the capabilities of women. "I was just angry at the notion that
women should be Victorians laced into corsets and prevented from
doing intense exercise. And that the medical community, or the sports
medicine community, was expressing that social limitation of women's
lives as medical problems in cross sectional, prejudicial studies." According to Prior, many studies that intended to look at the effects of
energetics on the menstrual cycle compared groups of women that dif-
fered along several dimensions, not merely how much they exercised.
Early research collected data from sedentary suburban women and
compared it to women at their first marathon training camp, confounding
lifetime differences in psychosocial stress, nutrition, and energetic constraint. Prior decided to design a study that made the playing field equal
in terms of reproductive maturation, observing women who were already runners, as well as those who were normally active or sedentary,

over time. Prior's research, which has held up over the past few decades, found that recreational or amateur exercise tends not to have significant effects on the menstrual cycle unless the athletes are undereating.[20] This suggests that many recreational exercisers are not putting their bodies in an energetic deficit, which is what we know is needed to appreciably affect the menstrual cycle. Menstrual cycle researchers have shown that the ovaries respond in a graded manner to stressors, particularly energetic constraint. Across many studies in many different populations, there is a well-established pattern by which the menstrual cycle changes when you have less and less energy left over after maintenance to contribute to reproduction.[21]

When your current energy availability is high and you have a lifetime of being well nourished, your menstrual cycle is as close as it can get to Norma's: there will be variation, but you will have a higher proportion of ovulatory cycles, and those ovulatory cycles will have a big midcycle rise of estradiol and a big rise of progesterone in the second half of the cycle, or luteal phase. Your cycle is dependent on both past and present access to resources to set your typical range of ovarian hormone concentrations. Those who experienced childhood or adolescent energetic constraint for a variety of reasons—helping their families on the farm, playing sports, or more extreme events, like experiencing famine—will have a menstrual cycle that looks a little different, even if their current energy availability is high.

At this first stage of energetic constraint (or a childhood environment of energetic constraint), changes occur in the luteal phase: lower progesterone concentrations, a slightly shorter phase, or both. With increasing energetic constraint, differences also appear in the first half of the cycle, ranging from lower estradiol to a longer follicular phase. Next, you may stop ovulating (anovulatory cycles), which for different people can cause both shorter and longer cycles. More constraint and you may experience a pretty irregular period (what is sometimes called oligomenorrhea). At the most extreme end of energetic constraint, people stop cycling and menstruating altogether (called amenorrhea).

These changes, which are primarily internal, can appear to cancel each other out. For example, if you have a shortened luteal phase and a

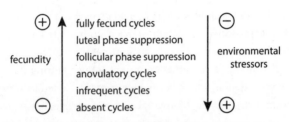

The progressive alterations of ovarian function as first proposed by Prior (1985) and Ellison (1990).

lengthened follicular phase, your menstrual cycle will still be your usual length. Similarly, even if you stop ovulating, you may experience no major changes in menstrual cycle length or the number of days you menstruate. For this reason, anthropologists and other scientists interested in human biological variation like to collect saliva, urine, and other bodily fluids to measure hormone concentrations throughout the menstrual cycle. By combining our knowledge of a person's hormone concentrations, their time of ovulation, their childhood environment, their own lived experiences, and the kinds of stressors they are facing, we can begin to understand how an individual's ovarian function is responding to the stressors in their life.

Experiencing various levels of energetic constraint from time to time should not be viewed as a bad thing. Some luteal and follicular suppression and the occasional anovulatory cycle are just part of the fluctuations of life. These suppressions are also reversible: I track my own menstrual cycle and have the occasional short cycle that tends to signal a shortened luteal phase—twenty-five to twenty-six days for me. When this happens, it is a signal that I might be undereating or overtraining; when I adjust for the next cycle, I'm back in the twenty-eight-to-thirty-three-day range. Too far the other way—thirty-four or thirty-five days, a far rarer event for me—and I suspect, but have not bothered to confirm, that I am not ovulating. Since I am not trying to get pregnant, the flexible responsiveness of my body to avoid putting resources toward reproduction when I just don't have them is totally fine.

Moving between these different menstrual cycle patterns is part of the normal ebb and flow of life. It means that the system is working correctly—that your body is good at shorting your reproductive effort in order to give more to maintenance. In other words, your body is prioritizing your survival over your reproduction, as it should. Prior herself has shown that, in a group of almost four thousand healthy Norwegian women, about a third of them were not ovulating when sampled. Rather than a cause for concern, this result shows us how common it is to move up and down the spectrum of ovarian function and that we should not expect to experience a fully ovulatory menstrual cycle that is exactly twenty-eight days long, every cycle, from menarche to menopause. Rather, substantial variation based on movement, training load, nutrition, and other stressors creates a variety of patterns common to the experience of menstrual cycling. Indeed, we still have a great deal to learn about the effects of more variable forms of exercise, let alone the interactions between these different factors.

Most research on physical activity and menstrual cycles has focused on urban, white women who perform largely aerobic forms of exercise (for example, running, dancing, cycling). We know next to nothing about how other forms of physical activity affect the menstrual cycle. The few existing studies on resistance training (for example, Olympic weight lifting or CrossFit) were performed with tiny sample sizes, groups as small as six to eight women. What little we do know suggests that resistance training and acute bouts of exercise (like high-intensity intervals) both seem to cause short-term increases in estradiol that, along with often short-term increases in testosterone, help with the tissue remodeling involved in building muscle.[22] Some older work hypothesizes that with enough of these bouts of increased estradiol, one can initiate the HPO axis negative feedback loop (discussed in chapter 2) and suppress ovarian function.[23]

Beyond exercise, there are so many other forms of movement! While many populations across the planet do not exercise, they still perform plenty of physical activity. The farmers that I study in southern Poland have until very recently farmed by hand. During harvest season, this involves cutting grain with a scythe, laying it out to dry, and gathering it to process or turn into bales of hay. Polish women and men in this population perform roughly equal work in the fields, but the women also pound grain, churn butter, beat rugs, hang laundry, wash windows, prepare dinner, and care for children. The director of my field site, Dr. Grazyna Jasienska, has shown that this moderate physical activity corresponds to lower progesterone concentrations, both at the population level and when comparing Polish women during their more intensive harvest season to the same women during the winter months.[24] Our work has also shown an older age at menarche, and shortened luteal phases, to correspond with higher everyday physical activity.

Dr. Katharine Lee has taken this work a step further with a rather important feminist question: Why do we pay so much attention to exercise and traditionally male movements and so rarely pay attention to the everyday domestic labor that takes up the days of most women of the world? Using Fitbits, the popular activity trackers, she has shown that Polish women take more steps than Polish American women and that they are lightly active for far more of their day. She has also found that rural Polish women have off-the-charts bone mineral density in their wrists (literally, they are several standard deviations above the reference values), likely because of the manual labor they do.[25] You don't need to run for ten miles for your activity to matter to your body: the fact that Polish women are lightly active rather than sedentary much of the day is probably a big reason why they tend to have lower average progesterone concentrations and shorter luteal phases, lower breast cancer risk, and lower bone fracture risk.[26]

One might wonder if rural Polish women, given their lower average progesterone concentrations, have fertility troubles. As far as we can tell, they do not, at least not any more than any other population. In our samples, research participants didn't suddenly start up a regimen of harvest season–level farmwork coupled with intensive domestic labor. Rather, this is work they have been doing in some form since they were children. In every population anthropologists have sampled so far, people who have a long history of physical activity tend to have lower average hormone concentrations. If you were to compare a southern Polish farmer to a third-generation Polish American from Chicago, much of the time you'd find that the farmer had a later age at menarche, lower progesterone, and possibly even a shorter luteal phase. But rather than indicate ovarian suppression and a less fecund cycle, these numbers represent what is typical for their population. If you need any more proof: one of the villages we work in is the most fertile village in Poland, and many of the families have up to a dozen children. The summer I lived there, I was told again and again of the year that Pope John Paul II blessed the village because of their good Catholic commitment to fertility.

At the end of the day, physical activity is meaningful for mental health, bone health, cardiovascular health, and several other aspects of our well-being. How much we move around, and how long we have been moving around, also matters to our menstrual cycles and fecundity. There is, however, no straightforward relationship between movement and menstruation—the range of variation that characterizes any given population shows the weakness of clinical thresholds when we all lead such different lives.

UNDEREATING AND UNDERCOMPENSATING FOR ACTIVITY

The energy expenditure from physical activity is usually compensated for by what we eat. Similarly, when people have less access to food, they often modify their activity to compensate for the shortfall.[27] There are, however, times that people do not or cannot compensate for physical activity with enough food. When I was a freshman in college, I ran track. I became fast friends with the other first years, taking classes together and eating meals as a pack. I rarely if ever missed practice but was among the weakest on the team—a walk-on triple jumper and hurdler on a team that had future Olympians in jumps and hurdles. Never in my life had I felt so untalented. These were extraordinary—and I should add, kind and humble and fun to be around—athletes. But perhaps, up against all that talent, I became just a little vulnerable. One of my other teammates offered nonstop talk of diets. Skinny as anyone else on this world-class team, she talked endlessly of how fat and slow she was and how she was going to completely stop eating. She talked about how much faster she would be if she just ate less.

The first year of college felt utterly out of control for me. In addition to my anxieties about my place on the track team, I also found that in many ways my high school education had left me far behind my prep school friends. On top of that, I often stayed up late doing homework and chatting with my friends and would find myself unable to wake up in time to get breakfast before my first classes of the day. I started skipping breakfast. It felt like a choice, like I was making a choice to control my diet to get better at track. I enjoyed the feeling of control I experienced from not eating in the morning. Next, I started skipping lunch. By the indoor track season, I often went to practice not having eaten anything yet that day.

Luckily, I eventually came to realize that these behaviors and habits were not healthy. Perhaps it had something to do with the fact that I completely tanked my pentathlon and heptathlon performances that year. Or maybe it was because, by the spring semester, I started performing better in my coursework and began to figure out my place at school. Somehow, I went back to eating on a regular schedule and have been fortunate to have never returned to that dangerous place again.

The messaging in women's sports is often that we need to be light and thin to perform at our best—even when the research directly contradicts these claims. Even when student athletes accept that they need adequate muscle to perform well at their sport, they often feel pressure to remain thin and "feminine."[28] Alongside this messaging, athletes can be told that if they lose their menstrual cycles, they are doing something wrong. In fact, these athletes are often warned about the female athlete triad: the relationship between menstrual irregularities (especially amenorrhea), a negative energy balance, and low bone mineral density. As a young athlete, I learned that I should get on the birth control pill to up my estrogen and decrease my risk of osteoporosis, so I have no idea if my disordered eating in my freshman year of college influenced my menstrual cycle.

Sports that have cultural associations with beauty and thinness—dancing, for instance—are those in which women have the highest risk of developing the female athlete triad. Dr. Prior has found that intense exercise influences the menstrual cycle, with the effects ranging from luteal suppression, to anovulation, to amenorrhea.[29] In some of Prior's earliest work, she documented case studies of marathon runners who had difficulty conceiving; in each case, the runner conceived after a six-week running hiatus.[30]

As subsequent researchers have observed, however, many of the runners studied were not merely exercising more than the norm. They were also quite lean and, in some cases, even had subclinical eating disorders. This raises the possibility that the primary problem was not exercise, per se, but rather the fact that women who compete in the sports most under the thrall of white patriarchal expectations of thinness are likely to develop eating disorders. Indeed, several important researchers, including Drs. Mary Jane De Souza and Nancy Williams, have made important strides in identifying disordered eating, not exercise, as the cause of the ill effects of the female athlete triad.[31] A growing number of health and fitness professionals, podcasters, bloggers, and influencers have also started discussing the ways in which calorie restriction can influence the menstrual cycle, particularly when someone is not eating enough to support their physical activity expenditures. And substantial historical research points to racist and sexist origins of fat phobia, which originated with the transatlantic slave trade and was strengthened by the rise of Protestant beliefs that to overeat was to be "ungodly." As a result, we developed a set of beliefs about a person's size that, as will become clear shortly, are entirely contradicted by scientific evidence.

The problem of undereating in elite training is perhaps nowhere more apparent than in the controversy over Nike running coach Alberto Salazar, who has come under fire from a number of his former athletes for abusive, body-shaming language that tied their performance to any perceived display of body fat.[32] Elite runner and track coach Amy Yoder Begley recounted to the *New York Times*, "If I had a bad workout on a Tuesday, [Salazar] would tell me I looked flabby and send me to get weighed. Then, three days later, I would have a great workout and he would say how lean I looked and tell me my husband was a lucky guy. I mean, really? My body changed in three days?" Mary Cain was at one time labeled the fastest girl in the United States for her outstanding performance as a high school runner. Upon joining Salazar's crew and the Nike training camp, the all-male leadership "became convinced that in order for me to get better, I had to become thinner . . . and thinner, and thinner." Cain lost her period for three years and broke five bones under Salazar's tutelage. Kara Goucher, who was also trained by Salazar and his cronies, disclosed that she had to sneak food into her room after the tiny meals allotted to her by an assistant coach.

Disordered eating isn't just for women, and by focusing only on them, we miss the harms experienced by all athletes. Ryan Hall, another American running phenom, struggled with weight issues and the belief that he had to be excessively thin to perform well. Yet in a recent book, Hall admits, "Looking back, I ran my very worst races when I was my lightest."[33] According to one of his coaches, his "emotional race weight" (which I take to mean the number that Hall thought made him race-ready) was 134 pounds on a five-foot-ten frame (though in other articles I've found descriptions of his race weight being closer to 127).[34] Historically, physical problems with elite male runners have been situated as *overtraining syndrome*, which they very may well be.[35] But it may also be that they were vastly undernourishing themselves to meet a perceived body image for the sport of endurance running. Since Ryan Hall retired and transitioned to weight training, he has put on forty pounds, most of it muscle. Yet Hall reports that despite this extra weight he is able to participate in speedwork with his wife, Sara Hall, whom he coaches, throwing up in one reported example a blazing 5:40 mile. While endurance runners may experience some benefits from being lighter, it's entirely possible that the body ideal of the "hollowed out cheeks of a marathoner" is not, in fact, ideal for the sport.

Researchers are recognizing that the female athlete triad is not reserved for girls and women. The phenomenon has been recently rebranded as RED-S: relative energy deficiency in sport. Like the female athlete triad, RED-S comes from not consuming enough food to compensate for the higher energy expended by many athletes, the consequences of which tend to include lower sex steroid hormones (estradiol, progesterone, and testosterone) and an increased risk of osteoporosis and fracture. The new consensus statement on RED-S also states undereating can cause impairments in cardiovascular health, immune function, protein synthesis, and metabolic rate.[36]

RED-S may not always occur from intentional undereating. Some studies have shown that exercising women, or women suddenly increasing their physical activity loads (when, for instance, they are training for their first marathon), do not always sufficiently increase their intake levels to match their expenditure. Now, if we aren't that great at assessing how much food we need, that should be an issue for Darwinian fitness. That is, if some people are less good at eating enough food to support their activity, those folks should not be able to allocate as much energy to their reproductive needs, and perhaps, in the long term, they may end up producing fewer offspring. However, before jumping to that conclusion we should again recognize that the types of people sampled for these studies are constrained: in some studies, this involves female endurance runners who may be invested in being thin; in other cases, studies have BMI restrictions on who is eligible to participate. It's not surprising to me that many of these participants are not eating enough or increasing energy intake when they increase energy expenditure. Rather than seeing a case of maladaptation—of inadequate hunger signals not allowing participants to keep up fecundity in the face of high expenditure—this may be a particular subsample of Westerners who exhibit low-level, subclinical, but culturally common food restriction.

Looking at a wider variety of populations, particularly those with subsistence behaviors focused primarily on foraging, pastoralism, or agriculture, we see that in times of food scarcity most people alleviate nutritional stress by changing their activity patterns. Agricultural populations move around less in winter, while forager populations that experience a hunger season reduce resting metabolic rates and in some cases manage work-arounds regarding food taboos to boost calorie consumption. This suggests that our bodies and minds are generally good at making the most of a bad situation. That is, in times of energetic stress, our bodies will do what they can to reduce needs, and our brains will tell us to reduce activity or acquire more food. General energetic stress from too few calories will lead to some ovarian suppression, as will consuming too little fat. But again, this is context dependent: we shouldn't expect a population that on average gets by on eighteen hundred calories a day of mostly fibrous carbohydrates to necessarily experience issues with fecundity because their diet doesn't have much fat. A person who switches to such a diet after being accustomed to more calories or more fat might experiences issues with fecundity as a result.

Across our evolutionary history and across plenty of modern populations, it is adaptive for our bodies to adjust our energy expenditure and intake as needed to do our best to maintain a neutral energy balance. It is also common, however, for this to be a challenge. People living through drought, famine, or seasonal food availability may from time to time reach a sufficiently negative energy balance that their ovarian function is somewhat suppressed. Resource allocation in the body can temporarily focus on maintenance, rather than reproduction; when access to food is restored, so is ovarian functionality. But many of us live in societies and cultures where food is either unavailable to some for long stretches—those living in food deserts or with food insecurity—or where people choose to refrain from eating enough to meet their energetic demands. A diet, we should recognize, is nothing more than an intentional relative energy deficiency.

Intentionally living in a state of negative energy balance for a long time is harmful not only to menstrual health but also to many other systems in our bodies. Yet people frequently choose to restrict their caloric intake because they mistakenly believe it is healthier to do so than to carry fat on their bodies.

AN ABUNDANCE OF ENERGY

There is, of course, a historical and cultural context to these choices. Not all cultures prize thinness and beauty standards, and even in Western cultures, full-figured but proportionate women were prized in written works and art throughout the Renaissance. As Dr. Sabrina Strings writes in *Fearing the Black Body: The Racial Origins of Fat Phobia*, beauty norms changed as white social elites sought to "create social distinctions between themselves and so-called greedy and fat racial Others." Early travelogues noted that many Africans were in fact slender. This perception shifted over time for several reasons, one being that white Europeans who had never traveled to Africa, nor met anyone with African heritage, wrote descriptive works about Black women. These white Europeans described Black African women as fatter, more gluttonous, and often more sexualized than white women. According to Strings, by the eighteenth century an "ascetic aesthetic" arose while "at the same time that gluttony and fatness were becoming associated with African women in scientific racial literature, the values of delicacy, discipline, and a slimmer physique were becoming associated with English women by the arbiters of taste and the purveyors of morality." This aesthetic, and the growing moral and medical beliefs that aristocratic white women should be pale skinned, slender, and fragile spread throughout the United States and the rest of Europe.

In the United States, the perception that thinness was a sign of morality and health (particularly in women) gained especially strong traction. As Strings shows, women's media from *Godey's Lady's Book* to *Cosmopolitan* were invested in highlighting only sources of beauty arising from Western and Northern Europe. The best American beauties were praised for their racial ancestry, as well as their "flower-like delicacy" and the fact that they were "tall and exquisitely slim." Toward the end of the nineteenth century, when public weighing scales became common, people also began focusing less on body proportions and more on absolute weights, as well as which were deemed more or less healthy. This focus became even more common in the early twentieth century when people began owning scales in their own homes.

When medical doctors began to study how weight influences health, some did push against the ascetic aesthetic, though it was because they believed that white women adhering to these ideals were harming their entire race. Forever since, women have been caught between aesthetic ideals of thinness and medical ideals of health. I can't help but wonder if this tension is part of where our fixation with being "toned" comes from—thin but not "too" thin, an impossible ideal of low body fat and some muscle development that cannot be easily maintained in a neutral energy balance way.

Despite fat phobia's racial and gendered pedigree, clinicians and researchers continue to devote an awful lot of time to the idea that too much fat on our bodies will disrupt our menstrual cycles. For the most part, spare energy, which is more common across many Western environments, is going to take the form of adipose tissue, or body fat. But different types of fat do different things: fat can be very good for you, and how much fat you carry does not correspond nearly as closely to health as many people believe. (I should also point out that health is not some precondition for being treated with dignity.) Indeed, fat stigma is proving to be a greater contributor to health problems than fat itself.[37]

Fat serves as an insulator, as an energy source for us (and for our babies if we are bodyfeeding parents), and as a part of the endocrinological and inflammatory systems. For example, fat makes estriol, a type of estrogen, as well as interleukin-6, a pro-inflammatory cytokine—both substances that are important to the menstrual cycle. The menstrual cycle also has important effects on the amount of fat on our bodies, as estrogens both promote the breakdown of adipose tissue and reduce its formation.[38] While folks who live in well-nourished populations tend to have the highest concentrations of estrogen *and* the highest percentages of adipose tissue, when you look at within-population variation in estrogen and fat among these women, a different story emerges. In one study, my friend and colleague Dr. Anna Ziomkiewicz compared the body fat and estradiol levels among Polish women with BMIs under twenty-nine to understand if the relationship between the two changed with body size. Rather than finding a linear relationship—where body fat increases along with estradiol—she found more of a U-shaped distribution. That is, within this range of weight for height, those women at the thinnest and fattest ends of the spectrum had the lowest estradiol concentrations.[39]

More recently, a sample of Americans were studied to understand the relationship between central and overall adiposity and estradiol and progesterone concentrations.[40] This sample included a broader range of BMIs, including those categorized as "obese" (a category that medically pathologizes people despite the fact that the category is itself a total invention). Among these participants, overall obese women had lower progesterone and higher estradiol. However, those with greater central adiposity had lower estradiol across the sample. The authors also looked at gonadotropins luteinizing hormone (LH) and follicle-stimulating hormone (FSH), which you may remember are important to ovulation as well as estradiol concentrations, and found that fatter women had lower LH and FSH overall but had higher peaks of these hormones than thinner women. There are two interesting conclusions to draw from these findings. First, it's not just how much fat you have on your body but where you store that fat that influences your hormones. And second, the body is quite adept at compensating for the effects of adiposity by ramping up gonadotropin amplitudes. Different bodies can use different pathways, and both can still achieve ovulation.

Fat phobia and fat discrimination have led us to assume not only that fat bodies are less healthy than thin bodies but also that when something goes wrong inside of a fat body, it must be because it is fat.[41] For instance, one of the classic papers that demonstrates a link between obesity and anovulation showed that women who weighed 120 percent or more of their "ideal" weight based on their height were more than twice as likely to have an anovulatory cycle.[42] But you know who was at an almost *fivefold* risk of anovulation in the same sample? The women who weighed 85 percent or less of their ideal weight for height! The link between obesity and anovulation is weaker than the link between undernourishment and anovulation.

Fat people are also often assumed to have pathological ovaries in the study of polycystic ovary syndrome, or PCOS, which is typically identified by the presence of at least two of the three following conditions: androgen excess (which can be characterized by measuring someone's testosterone, as well as characterized by someone's appearance, as with hirsutism beyond what is expected for that person's ethnicity and gender), less frequent menstrual cycles or anovulation, and polycystic ovarian morphology as visualized on ultrasound. Polycystic ovaries often look like they have a pearl necklace, as they have a number of large but unerupted follicles at the ovary's surface. (Note that, according to these criteria, a person with hirsutism and anovulation can be diagnosed with PCOS even though they do not have polycystic ovaries!) Several years back I had an ultrasound that indicated about ten unerupted follicles on each ovary, and at the time I only got my period every other month. I did not count as having PCOS because of my ten rather than twelve follicles.

PCOS is linked to fatness in well-nourished populations with higher rates of sedentary behavior. But as the old saying goes, correlation does not equal causation. Recent research suggests that there may be some confounding variable, causing both more PCOS and more fat in certain populations, rather than the energy storage itself being the problem. Both Indian and Iranian populations appear to have a high prevalence of PCOS despite population averages for fat that are lower than in the United States.[43] Among East Asian samples, women with PCOS on average have a higher BMI, but the average BMI for someone with PCOS in those samples is only twenty-two (well below the overweight and obese ranges).[44] The most likely culprit is insulin resistance. The body needs insulin to help it take glucose up from the bloodstream; when the body's tissues struggle to take in insulin, they can't take in as much glucose. The body can compensate by pumping out more and more insulin to account for the problem, but this will also contribute to insulin resistance. As a result, glucose remains in higher concentration in the bloodstream, which can lead to negative health outcomes, including diabetes. Insulin resistance itself occurs from a combination of genetics, stress, an energy surplus, and a higher carbohydrate diet. While there is some connection between obesity and PCOS, it is not as definitive or direct as we might have thought. In the end, fat phobia may be overblowing some of the effects of fat storage on the menstrual cycle while encouraging us to overlook the fact that reducing energy storage—creating an energy deficiency through a self-imposed or doctor-imposed diet—definitely reduces the functionality of the ovaries.

When fat people encounter fertility issues, they are often encouraged to lose weight. Patients will adopt severe calorie-restricted diets, take appetite suppressants, or have bariatric surgery, all to hit the BMI criterion set by their infertility doctor or clinic. A recent review of the effects of weight loss among obese infertility patients is especially troubling. Dr. Richard Legro of Penn State University College of Medicine describes study after study in which people seeking pregnancy are put on varying regimes of weight loss, only to end up with minimal to no improvement in their chances of getting pregnant.[45] No matter the outcome measured—maternal or infant morbidity, number of conceptions, or number of live births—the results are disappointing to those who think weight loss is some sort of panacea. And as Legro points out, the most desired outcome of people seeking infertility help, the birth of a healthy infant, is the one metric researchers rarely bother to measure.

In several studies, the weight-loss group had worse rates of fertility than the control group. In others, the weight-loss group experienced dangerous complications: higher risks of birthing smaller infants, pre-term birth, and even stillbirth and neonatal death. As Legro observes, "Never has so much time, effort, and money been directed toward so many obese pregnancies with so little public health benefit."

Fatness is not a health problem per se but a correlate of other system-wide phenomena. While an abundance of energy can in some cases contribute to disease or disease severity, a society and medical system biased toward valorizing thinness and shaming fatness also appears to be one that understates the harms of energy deficiency and overstates those of energy surplus. As we will soon discover, the psychosocial stress induced by stigma, discrimination, and abuse can have its own negative effects on the menstrual cycle. However, before we turn to the effects of psychosocial stressors, we will first examine immune stressors. In con-trast to energetic constraint, which literally reduces the energy you can allocate toward maintenance, growth, and reproduction, these two other stressors can complicate how resources are handled by the body. Because the mechanism by which our bodies react to immunological and psycho-social stressors is often inflammatory, the effect on the menstrual cycle is less about constraint and more about adding noise to the signal. In chapter 4, we'll discuss the immune stressors as a way to help explain what happens to the menstrual cycle when you get sick and also to start to understand how these inflammatory processes are involved in re-sponding to both immune and psychosocial stressors.

CHAPTER 4

Inflamed Cycles

ON MONDAY, February 1, 2021, Dr. Katie Lee sent me a message on our lab Slack that she wanted to "pique [my] curiosity regarding inflammatory response/vaccination/periods." Katie had been talking with other friends who had been vaccinated early. Those with menstrual cycles were having terrible periods while those on IUDs were having unexpected breakthrough bleeding. She and I exchanged a few thoughts about the subject, and I downloaded a few articles to review. But like many projects during the pandemic, it quickly moved to the back of my mind, crowded out by the responsibilities that lay directly in front of me—my classes, my administrative duties, my trainees, and my immediate paper drafts and revisions—as well as my own anxieties about getting infected with SARS-CoV-2.

All of this changed two weeks later when I became eligible for the vaccine. After receiving my first dose—a modern miracle of science—my period started earlier than usual, and I bled so heavily that I was swapping out overnight-strength pads every hour or so. Unlike the darker menses to which I was accustomed, this was bright red with little visible endometrial tissue. I was also incredibly fatigued. After about a day and a half of this, on February 24, I decided to tweet out a query: Has anyone else had changes in their periods since receiving the vaccine?

The response was unlike anything I've ever experienced on Twitter. I have had threads and posts receive attention before—including one thread in which I divulged the awfulness of my postpartum recovery after my second child—but nothing like this.[1] I was bombarded with commiserating responses, along with a smattering of people accusing me of fearmongering. Within a few hours, Katie and I were furiously messaging each other.

We thought it would be a fun, manageable project with what we hoped would be an interesting and useful outcome: validate a side effect and reduce fear that it was related to fertility. We would characterize some of these menstrual experiences—just as people had already been doing with sore arms, fatigue, headache, body aches, and fever—and maybe even help out an undergraduate who needed data for an honors thesis. Katie had the first draft of the survey done before dinnertime, and just a month later—after receiving the go-ahead from the Office for the Protection of Research Subjects at the University of Illinois—we were ready to recruit.

In our application, we noted that we expected about five hundred people to participate. We hit five hundred within the first few hours. Then six hundred, then seven hundred. By the next day, we were over one thousand. We received twenty-two thousand submissions during our first week and more than eighty thousand during the first month. We also received hundreds of messages from people who were thanking us for doing this work, sharing additional experiences, and asking us if something they were experiencing was dangerous (we sent a lot of responses telling people to call their doctors), as well as several media requests a day.

The interest in our research was in many ways thrilling. We were all going through this shared experience of a pandemic, but now there was a major pivot toward hopefulness as more and more people were getting vaccinated. Things were dire, but maybe, eventually, they would be better. Our project arrived during a moment when people were especially curious about their bodies and how their bodies might be affected by this new vaccine platform. Moreover, there was nothing far-fetched about the idea that vaccines might affect people's menstrual cycles. We have long known that immune responses can influence menstrual cycles and that vaccines can occasionally affect bleeding patterns.[2]

Nonetheless, many menstruating, formerly menstruating, and post-menopausal people shared with us that their family members laughed off their experiences; even worse, their doctors and medical personnel were dismissive and rude. Journalists who interviewed us for their stories also interviewed medical doctors, who often demonstrated an almost total lack of even the most basic knowledge about menstrual cycles. Medical doctors tended to assume that those experiencing heavy or breakthrough bleeding were imagining these changes because they were so stressed from the pandemic or that the entire line of inquiry was ridiculous because there was no biological mechanism that could possibly link vaccine activation and menstrual changes.[3] Most of the high-profile coverage of our research foregrounded pieces with these views, which served to cast us as ridiculous ladies feeling our ridiculous feelings about things that were definitely not real.

To some extent I'm used to this. The lab is centered on feminism and reproductive justice, and we explore topics that are stigmatized all the time—I often joke that we are in the business of making people uncomfortable. What's more, other scholars often assume that because we are anthropologists, we lack the relevant biological expertise to study what we study. While we use methods that are different than the big, heavily funded labs and programs, I would argue that for certain research questions our methods are better. Our work bears witness to lived experience, explores relationships, and develops hypotheses, which, if necessary, can later be leveraged into the larger-scale randomized controlled trials so loved by clinical and public health. As I mentioned at the outset of this book, in our lab we perform what anthropologist Dr. Anna Tsing calls the "art of noticing." Noticing allows us to see power, to follow multiple threads without allowing one to dominate. It is an essential scientific practice.

Katie and I noticed changes in ourselves. We noticed changes in friends and acquaintances and strangers. This noticing came from having bodies that menstruated and from our scientific knowledge of menstruation. This noticing also came, however, from adopting a feminist practice and sense of responsibility to notice what some with clinical and public health backgrounds were obstinately refusing to notice, even when right under their noses. Many feared that if we did not use only a specific set of methods (control groups! randomized trials!), we might learn the wrong lessons—or worse, that our work might encourage vaccine hesitancy.

These fears come from a certain rigidity, a belief that noticing what typically goes unnoticed will disrupt the dominant melody. This is, in part, the point of our research: noticing exposes unheard songs; noticing exposes complexity and the fact that most questions do not have single, tidy answers. And the overwhelming sense that I've gotten as I've been writing this book is that the cultural factors that have occluded our viewpoints and dampened our awareness of this complexity reside in the eugenic history of gynecology.

As I mentioned at the start of this book, the discipline of gynecology was professionalized and gained traction in the nineteenth century, in part, by removing the last vestiges of power held by Black and brown midwives. The paternalism that dominated the field resulted in only two lines of harmony if you had a uterus: either you had to be protected at all costs or your symptoms were a figment of your imagination. These harmonies create control and define the experiences of people with uteruses. I am asking in this chapter that we notice the polyphony of many more lines of music.

Noticing is especially useful when it comes to the reproductive-immune axis because listening longer helps us hear so much more complexity. For instance, it is likely that the postvaccine experiences of so many people—of heavy bleeding, clots, breakthrough bleeding, hot flashes, and extreme breast pain—are real *and* that these effects do not mean we have to have our fertility shielded from the vaccine. More broadly, the human reproductive-immune axis has a lot of dissonance beyond the possible effects of vaccines on periods, if you listen: our bodies have evolved to create conditions where sperm can survive to fuse with an egg while also trying to prevent other pathogens from sneaking in at the same time. If our immune response is too aggressive, pregnancy doesn't take; if it isn't aggressive enough, the vaginal tract or even uterus can experience infection. Uteruses are robust and powerful: they regenerate tissue and harbor fetal materials that eventually can become viable life.

They literally make humans. The dominant paternalistic harmony simply does not fit the evidence.

This chapter, then, is about the complex relationship between our immune and reproductive systems—what we notice when we relax the desire to control women and gender-diverse people and dig in a bit more to our power. I will start by sharing how inflammation is crucial to the tissue remodeling and signaling that happens in the ovaries and endometrium, thus showing how these processes are intertwined (rather than at odds). Then I will discuss the concept of *menstrual hygiene management* (MHM)—what is or isn't hygienic, and who or what is managing it? Is there such a thing as unhygienic menstruation? Does it cause physiological harm? Finally, I will return to our vaccine project and what we've found. Some threads to notice relate to capitalism and colonialism; gender role anxiety and menstrual stigma; the ways in which inflammation is neither good nor bad. We can take the time to listen, and listen again, to the polyphonies of pro- and anti-inflammatory cytokines as they work their way through our bodies, interact with endometrial tissue remodeling and ovarian hormones, and produce intertwined melodies.

GENDER DIFFERENCES IN IMMUNE FUNCTION

One of the places where the paternalism of gynecology rears its head is in the way we describe gender differences in immune function. And while there are differences that have to do with gender/sex, disentangling them from chromosomes, hormones, lived experiences, misdiagnoses, and oppression is nearly impossible. Much of the immune system operates the same way for most people: when you are wounded, your body sets several processes in motion. These include inflammatory processes intended to stave off attack from an outside pathogen and regenerative processes such as clearing cellular debris, promoting cell differentiation, and stimulating growth factor secretions. These processes prioritize getting wounds cleaned up and closed up to protect the inside of the body from the bogeys (or in my children's case, boogers) on the outside. To orchestrate this task, the body must first stimulate inflammation and then eventually tamp it down—a generalized *crescendo* and *decrescendo* across all melodies.

Exposure to a pathogen stimulates a similar process. The innate immune system—that which is activated at wound sites and microbial exposures—reacts to the molecular patterns of microbes quickly but generally. These general defenses give the adaptive immune system time to develop more specific defenses after the body has identified the potential invader. From this point, the body's processes start to become more specific: one melody might crescendo and lead to a new and specific variation on the theme while other melodies grow quiet.

This is, at least, the story of acute immune stressors and responses. Chronic immune stressors, which operate somewhat differently, come in many forms, but globally the most common are caused by parasites. Many critters have evolved to be parasitic in humans—from protozoa (single-celled organisms like amoebas) to helminths (worms!) to ectoparasites (those that live on the outside of our bodies, like lice). In parts of the world where people do not have access to running water or working bathrooms, they are often exposed to parasites in human and animal waste. Since humans evolved alongside our parasites, our immune systems evolved with the capacity to handle a certain level of pathogen burden. If we hadn't responded adaptively and flexibly to this constant immune challenge, we would have been sicker and less able to make, let alone care for, babies.

Nowadays, many of us have access to clean bathrooms and running water. We also spend less time outdoors, which reduces exposure to soil microbes (from playing in the dirt, foraging, gardening, or farming) and zoonoses (from hunted or farmed animals) while increasing our exposure to air microbes and air pollution. While vaginal childbirth and breastfeeding help subject us to our birthing parent's microbiome and thus to more of these benign exposures, cultural factors, systemic pressures for those who work outside the home, and the medicalization of labor mean these are not easy or available options for everyone.[4]

While there are many positive consequences to this change in exposures, those of us with fewer exposures to parasites and microbes often have immune systems that tend to get bored. This is akin to the situation my family has with our collie Archie. Archie is a smart herding dog who, despite our best efforts, spends too many hours at home. Without sheep to chase around and keep him occupied, Archie often engages in destructive behaviors, like eating piano music, tissue boxes, and mail (our mail carrier has stopped putting it through the mail slot, opting to place our letters and magazines against the door). Without a way to fulfill his preferred purpose, Archie finds other things to do, and these are typically things we'd really rather he lay off doing.

The idea that a bored immune system would wreck our bodies like an understimulated herding dog was first called the hygiene hypothesis, but more recent updates prefer terms like the old friends hypothesis or the targeted hygiene hypothesis (both, in different ways, emphasize our reduced exposure to friendly microbes).[5] This hypothesis suggests that giving our immune system something to do keeps it from attacking non-invading materials, ranging from benign proteins found in food (think of peanuts, and therefore peanut allergies) to pollen (think of hay fever) to our own bodies (for example, autoimmune diseases like multiple sclerosis or lupus). Many infectious diseases are still unwanted and dangerous— in no way do I recommend that you stop washing your hands or your produce or stop masking in cold and flu season or during a pandemic— but, according to this hypothesis, our change to a more sanitized environment has had benign *and* harmful consequences.

How does this big-picture (and admittedly simplified) story relate to the effect of immunological stressors on the menstrual cycle? It has to do with the dominant melody around what are called Th1 and Th2 systems. Th1 and Th2 are classes of cytokines, or inflammatory signals, derived from T lymphocytes. The Th1 response, which is largely pro-inflammatory, is responsible for the cell-mediated immune response, meaning it kills the tiny parasites. The Th1 response tells macrophages to target bacteria and protozoa that shouldn't be in our system. The Th2 response is part of the humoral immune response, meaning it kills helminths (worms!). The Th1/Th2 balancing act is a major part of the scientific narrative when it comes to sex differences in immune function. According to the standard narrative, boys have a stronger Th1 response and girls a stronger Th2 response. The immune systems of girls and

women are further complicated by the need to kill intruders like helminths while not killing embryos—imagine how easy it might be to
mistake one invader for the other.

But there aren't just Th1 and Th2 responses; there are also Th17 and
possibly also Th0 responses. And like most sex differences, there is so
much overlap in the ranges of variation that it makes more sense to
observe our similarities rather than our differences.[6] As the story has
become more complicated and, I suspect, as science has become a more
diverse enterprise, we have no longer been able to stick to the standard
immunological narrative. The point is not that there are no differences
between sexes or genders but rather that the idea of a dichotomous immune system is not a story that serves us any better than it would to be
ignorant of gender in our approach to health and disease. Women do
tend to have more immune conditions than men. About 8 percent of
people in the United States have autoimmune disease, but more than
three-quarters of those people are women.[7] This statistic is consistent
with the idea that men have stronger Th1 responses compared to
women.

Perhaps, however, rather than identify a binarized sex difference,
what we need to pay attention to are the dominant sex steroids in a given
person. In other words, people who are or have been more estrogen
dominant might be more likely to have one type of response, and people
who are or have been more testosterone dominant might have another.
The prevalence of allergy and asthma changes that occur after puberty
lends some credence to this idea: in childhood, asthma and allergies have
a higher prevalence among kids with testes. However, once kids hit
puberty the ones with ovaries start to outpace the ones with testes in
the incidence of both of these conditions.[8] This is, of course, also the
time that many testes tend to produce higher quantities of testosterone
and that ovaries produce higher quantities of estrogen and progesterone. And while I have found little research on immune function in non-
cis people, I did find two case studies of transgender women developing
lupus after estrogen therapy, cited in a review on hormone therapy.[9]

If aspects of immune function are influenced by sex steroid hormones, this would also exacerbate the hormone-mediated difference in

autoimmune diseases in industrialized countries since both estradiol and testosterone tend to be substantially higher in folks living in these environments. It's as though the industrialized environment is creating a difference between those folks who are testosterone or estradiol dominant that, under other conditions, might not be that big or meaningful. Perhaps, then, it's not that girls are naturally oversensitive to immune challenges but rather that hormone-mediated differences in immune function are enhanced in industrialized environments, which are vastly oversampled. As we discussed in chapter 3, energy availability through one's lifetime in part determines the typical range of adult hormone concentrations: this creates significant and meaningful differences between populations as well as between individuals living in the same culture, country, or geographic region. What's more, some of the basic mechanisms of immune function share pathways with the basics of ovarian function, which adds complexity to the menstrual cycle.

INFLAMMATION AND DISTRIBUTING RESOURCES TO THE OVARIES

Recall that the resources allocated to maintenance effort, growth, and reproductive function are dependent on whether you are pre- or post-pubertal, pre- or postmenopausal, how much energy you put into the system in the form of food, and how much you constrain it with physical activity. In times of plenty, allocation is an easy matter, and everyone gets what they want. But in times of constraint, our body must decide who has to be shorted. Typically, we short reproduction first if it is one of the systems to which we are allocating, especially if there are years yet before our reproductive span is over. Immune function is part of maintenance effort, so immune challenges may require that we allocate more to maintenance, at the expense of growth and reproduction.

However, immune function has important effects on our use of energetic resources beyond resource allocation. Because the communication practices of immune function are inflammatory and so are many of the functions of tissue remodeling, a big draw toward immune function might disturb our body's distributions of energy. Both growth and reproduction require constant tissue remodeling, which means they use the same communication pathways as does immune function. In these ways, inflammatory signals therefore can draw energetic resources, apply noise to the signal, or make the process less efficient. Inflammatory processes are necessary and important to just about every process in our bodies, so when they are invoked in a big immune response, it's like they are speaking over all of their other uses. Immune function becomes the dominant melody, and other inflammatory processes become, for a time, harder to hear.

Inflammatory signals coordinate acute elements of the immune response to begin the processes of healing and repair or to kill off infection. The two inflammatory biomarkers that have received the most attention on this topic are C-reactive protein (CRP) and interleukin-6 (IL-6). CRP, made in the liver and in fat cells, is often measured as a signal for systemic inflammation because it changes in response to things like illness and pathogen burden, as well as changes in energetics (including how much fat you have, given that fat is a source of CRP). It also plays an important role in cardiovascular disease as well as host pathogen defense systems.[10] When the CRP molecule breaks down into its individual parts, its main job is to orchestrate processes of inflammation and tissue remodeling. CRP is a boss at both recognizing infection and activating the immune processes that clear out foreign particles from our bodies.[11]

IL-6, which largely rises and falls with CRP, is produced in many places, including bone, blood vessels, immune cells, the ovaries, and fat cells. It's pro-inflammatory, in the sense that it is involved in the acute phase response of the immune system, which means that, like CRP, it rises in response to an immune challenge. But it's also *anti*-inflammatory. For instance, when secreted by muscle during physical activity, it down-regulates other inflammatory markers.[12] IL-6 and CRP are both important to our understanding of chronic inflammation as well.

The ovaries and uterus undertake significant tissue remodeling over the course of the menstrual cycle, which changes in CRP and IL-6 often reflect. While the ovaries manage the overall process of menstruation, they themselves also undergo substantial changes during each menstrual cycle. First, as we learned in chapter 2, they are continuously recruiting waves of follicles to develop in the effort to select a quality dominant follicle. These small organs—in premenopausal folks they are three to five centimeters long—house millions of tiny oocytes that grow, regress, and undergo selection two or three times every menstrual cycle.

During ovulation, the extracellular matrix (the collagens and other materials that hold together the three-dimensional structure of cells) at the follicular wall degrades and thins to the point that the dominant follicle can rupture its surface and emerge. The ovary then must tend to the repair of this small wound, as well as to the maintenance of the corpus luteum, the "yellow body" left behind after ovulation that produces progesterone and maintains the endometrium. All of these remodeling processes involve inflammation.

When we started looking at the relationship between CRP and ovarian function in our lab with Drs. Angela Baerwald and Roger Pierson, we found that among urban Canadian women, CRP is higher for those with three follicular waves, compared to those with two. CRP is also higher for those with more major waves—that is, waves with a dominant follicle. Since more waves indicate the need for more tissue remodeling, this was an expected (though cool) finding.[13]

The endometrium is also a site of extraordinary tissue remodeling. During the early stage of one's cycle, the uterus is busy making spiral-shaped arteries, growing tissue and differentiating it into those nice nooks and crannies so beloved by embryos. Progesterone withdrawal toward the end of the cycle stimulates tissue breakdown for menstruation. Then, when menstruation does initiate, endometrial tissue swells, and immune cells flood the area.[14] The inflammatory effects of menstruation played an important role when we decided to take a closer look at the CRP and follicular waves data. The neat thing about Roger and Angie's data set is that unlike most clinical samples, which have at best two to three serum samples to measure over a menstrual cycle, their sample included blood that they had collected every three days. As a result, we could look at the daily ultrasounds they had collected on ovarian function and follicle size alongside longitudinal sampling of CRP from blood. This allowed us to determine whether women with three waves had elevated CRP throughout the cycle or whether this increase was tied to any other potentially inflammatory event during the menstrual cycle. We determined that CRP in women with three versus two waves didn't have any discernible pattern difference . . . except that there seemed to be a big spike at menstruation. And when we looked at the data more deeply—comparing not just women with two versus three waves but those who had different patterns of minor waves (without a dominant follicle) versus major waves (with a dominant follicle)—we found something weird. First, among women with three major waves, the second wave seemed to co-occur with menstruation. And second, almost all of the difference between three- and two-wave women could be explained by the giant spike in CRP in women with three major waves from within that group. What we suspect is that, among women in our sample who had three major waves, the co-occurrence of two major inflammatory events—a follicular wave and menstruation—is what caused the spike in CRP.

Let me put this nerdy finding into the broader context: elevated CRP is used to diagnose people as prediabetic and as at risk for cardiovascular disease. Unless clinicians control for a patient's menstrual cycle, elevated CRP levels may lead to a misdiagnosis. We aren't the only researchers to demonstrate changes in CRP across the cycle: others have documented elevations at ovulation or menstruation depending on the sample.[15] So when we witness elevation in CRP levels, this may be because other factors are affecting ovarian or endometrial function and thus driving up or down tissue-remodeling processes that drive inflammation.

In the Canadian sample, collection protocols made it possible for me to look at CRP variation through the cycle and look at their association with processes like follicle waves or menstruation. This allowed us to ask whether there was an association between elevated CRP and unusually high inflammation moments like the co-occurrence of a major wave and menstruation, present only in a handful of participants.

This CRP was also from serum, and the primary CRP that you find in blood is its large, pentameric (five parts together) form. The job of CRP in this form is to recognize pathogens and damaged cells in the body; if serum CRP is high, these are likely tasks the body feels it needs to be doing, which could mean a major inflammatory challenge is happening. However, I had urine left over from my dissertation research in rural Poland (ah yes, living the dream), and I wanted to take this research one step further.

In our rural Polish sample, we don't have data on their follicular waves, but we do have daily estradiol and progesterone concentrations throughout participants' menstrual cycles, as well as reproductive history. This allows us to ask whether there are associations between ovarian hormones and CRP. This CRP is from urine, which means we are looking at the "broken-down" form of CRP because its bigger form cannot fit through the kidney's filtration system. We also happened to time the urine collections to the middle of the follicular and luteal phases, which would rule out the possibility that tissue remodeling from ovulation or menstruation might raise CRP levels. Therefore, urinary CRP tells us something about how our bodies orchestrate inflammation and tissue-remodeling processes, probably related to how the body is dealing with external stressors (for example, immune or psychosocial challenges), and our collection protocols unintentionally removed the "noise" caused by menstruation inflammation.

In the rural Polish sample, we found that women with high urinary CRP tended to have lower progesterone. When we controlled for indicators of early environment like age at menarche, women with high CRP also tended to have lower estradiol concentrations. Taking all the methodological differences into account, a negative association between urinary CRP and ovarian hormones could mean that external stressors can influence inflammation, which in turn has consequences for menstrual cycles. If elevated enough for long enough, inflammation could have consequences for fecundity as well.

So it goes both ways: CRP can tell us about tissue remodeling from within and external stressors from without, both of which generate inflammation. Many study designs can't distinguish between the two, which makes things even trickier. But this also points to an additional issue. Inflammatory markers do not only play a functional role within the body—they are also signals that help tell the body how to allocate resources to deal with stressors and thus have disruptive potential.

The final element of this story is that, as I've discussed in previous chapters, we find plenty of variation in what clinicians would call *normal spontaneous cycles*. This variation is part and parcel of a system that is at its best when it is responsive to its environment. This variation does not mean that anything is necessarily wrong, and for the most part, the effects that current perturbations have on a menstrual cycle affect just that cycle (and maybe the one after). Unless one is, at that exact moment, actively trying to have a kid, then fluctuations in hormones, inflammatory biomarkers, and menstrual flow have no bearing on that person's fertility. Even in a cycle where someone is trying to conceive, most of the variability discussed in this book would not appreciably affect the chance of conception. The hyperawareness with which we have been taught to monitor change in our menstrual cycles and the equation of this variation with harm to fertility has more to do with the paternalism of medicine and its ongoing issues with menstrual stigma than with menstrual cycle biology. I offer insights into this variation not to contribute to that hyperawareness but to explain how perhaps that hyperawareness does not serve fertility goals to the degree we think it does. I would even contend that hyperawareness of menstrual cleanliness is another method of control, rather than being tied in an appreciable way to our health or well-being.

MENSTRUAL HYGIENE MANAGEMENT

Menstrual stigma is immensely harmful to people who bleed. It keeps them from learning about their bodies, from communicating, and from being able to access the resources they need to have safe periods, not to mention safe sex and reproduction. Menstrual stigma creates limits around what people feel they can do when they have their periods, as well as what they can in fact do. When a person cannot access tampons or menstrual cups, it's not just that they "feel" they can't go swimming in the pool with friends; this lack of access operates as a structural limitation on movement, socialization, and enjoyment. As I pointed out in chapter 1, menstrual stigma isn't some sort of human universal—plenty of cultures have a neutral or even positive view of menstruation, including early Western European cultures. However, given that much of the scientific conversation about periods is dominated by Westerners and that even non-Western science has been colonized by Western perspectives, the dominant science tends to be influenced by and reproduce menstrual stigma.

The influence of menstrual stigma on science pervades public health and its interventions around issues of MHM and period poverty. When Western (mostly British or American) public health researchers look at gender disparities in health, finances, or education in non-Western communities, they increasingly turn to periods as a site of intervention. These projects generally aim to increase access to menstrual products and safe places to change and/or wash them: undoubtedly a very worthy goal. In some projects, the objective is simply to get disposable or reusable products in hands; in others, it is to create private or semiprivate places to change; in still others, it is to teach people who menstruate how to make their own products or even launch their own microbusiness making and selling products.

Dr. Chris Bobel, a professor of women and gender studies at the University of Massachusetts-Boston and the author of *The Managed Body: Developing Girls and Menstrual Health in the Global South*, offers what she calls an "invested critique" in the MHM approach.[16] As Bobel points out, these interventions seek to "manage" the period in such a way that it is invisible to others; the acceptable period, they imply, is "the one we don't know about."[17] Similar to Dr. Sharra Vostral's conception of menstrual products as "technologies of passing," Bobel describes a world where (mostly) Western people identify what they see as a problem among (mostly) non-Western people through their own cultural lens: it is the responsibility of menstruating people to manage and hide their periods in order to succeed in the world.[18] As with much of Western thinking, the body is the obstacle to a life of the mind, and our narratives of progress, productivity, and capitalism are deeply embedded in the ways in which we find it necessary to manage menstruation and menstrual symptoms.[19]

Menstrual stigma pervades the way we approach the issue of MHM; it even pervades the term itself. Why do we seek hygiene for menses?

The MHM literature suggests two reasons that managing periods will help resolve gender disparities. First, access to sanitary and safe menstrual products (and places to change them) is thought to increase educational attainment. The thought here is that girls miss out on educational opportunities because they do not have menstrual products and/or a safe place to change them at school. There is some truth to this: schools in nations that have experienced centuries of colonialism and extraction from the West are less likely to have sanitation facilities or safe places for menstruating people to change out menstrual products as needed. Drawing on the work of Dr. Max Liboiron, a geographer and director of the Civic Laboratory for Environmental Action Research at Memorial University, I understand colonialism as "a way to describe relationships characterized by conquest and genocide that grant colonialists and settlers 'ongoing state access to land and resources that contradictorily provide the material and spiritual sustenance of Indigenous societies on the one hand, and the foundation of colonial state-formation, settlement, and capitalist development on the other.'" As Liboiron stresses, "Colonialism is more than the intent, identities, heritages, and values of settlers and their ancestors. It's about genocide and access."[20] One of the things we need to reckon with in any anthropological book that tries to understand human biological variation, adaptation, or evolution is that the majority of that work is produced by white people like me, who think that because our research questions are interesting or important (to us), we should be considered a benign influence on the people whom we study.

On its face, it seems like an excellent idea to try to better the conditions for young girls in postcolonial countries to improve their educational and future financial attainment. However, the evidence that menstrual product access influences educational attainment is weak. The two largest studies on the topic to date, one a review and the other a cluster randomized controlled feasibility study, suggest that neither menstrual product access nor menstrual education play much of a role in school attendance or dropout risk.[21]

The second reason typically given for MHM projects is that adequate menstrual hygiene will reduce infectious disease risk. If menstruation is unhygienic, or at least one's menstrual management practices are unhygienic, perhaps that invites unwanted bacterial or viral guests to the scene. One sample of 181 adolescent girls in Karimnagar, India, found that "unsanitary cloth," defined as using a cloth, sometimes more than once, was associated with an increased risk of urinary tract infections compared to using a disposable sanitary pad. This study did not, however, control for other confounders such as socioeconomic status or safe water access, which are probably correlated with the use of cloth pads.[22] In another study of 558 women in Odisha, India, those who used reusable cloths and practiced less personal washing were more likely to have candida infections.[23] Several other practices, such as having to change menstrual products outside of a toilet facility or dry and store reusable products in places where they may not get fully dry, also increased the risk of bacterial vaginosis and candida infection. And yet a 2020 study of 240 adolescent girls in Kibogora, Rwanda, found no association between urinary tract infections and menstrual hygiene practices.[24]

Another study that casts doubt on this association is a 2016 cluster randomized controlled study of 644 adolescent girls in rural western Kenya. Randomized controlled trials are intervention studies where the aim is to reduce selection bias by randomly allocating participants to different treatment groups. A cluster randomized controlled trial is one in which preexisting groups—in this case, girls at thirty different schools—are randomly assigned the different treatments. Ten schools received reusable menstrual cups, ten schools received menstrual pads, and ten schools served as a control group where the girls' "usual practices" were observed. "Puberty and hygiene training" was given to all enrolled girls across treatment groups. The authors found that sexually transmitted infections decreased in the pad and cup treatment groups compared to the control group. However, despite analyzing that data in many different ways, reproductive tract infections were not all that different between groups. Only after more than nine months of cup use were bacterial vaginosis rates slightly lower than the rates in the pad and control groups.

But, note the connection between the incidence of sexually transmitted diseases and the availability of menstrual hygiene products. As it turns out, several studies have shown that menstruating children and adults in some settings trade sex for food, clothing, menstrual products, and school supplies. What these places have in common is high poverty and few means to be employed beyond their substantial and unpaid domestic labor.[25] Given these circumstances, MHM interventions may not help menstrual health directly but rather reduce participants' need to engage in transactional sex to acquire basic necessities. As a result, the participants were less frequently exposed to sexually transmitted infections. To summarize: the pads (or cups) aren't reducing infections. People are living in poverty conditions, often because of centuries of Western extractive practices, and Westerners are providing a literal bandage solution to a structural problem.

In the spirit of Bobel's "invested critique," I hope it is clear here that my intention is not to slam the purveyors of colonial "good ideas" but rather to notice from where they come.[26] I observe that many MHM papers have white Western women authors (usually but not always British or American) in the coveted first and last authorship positions, a problem that is pervasive across academic global health.[27] Collaborators who live and work in the countries under study are often sandwiched in the middle. It doesn't escape my notice that my papers on research conducted in Poland have Americans typically as first and last authors, with Polish authors sandwiched in the middle—this is not me pointing fingers (or if I am, I am also pointing one at myself). But the aforementioned studies on the connection between MHM interventions and infection risk are borne out of imbalanced collaboration, where the people with funding and power set the agenda, decide on the intervention, and specify the relevant outcome measurements.

Periods themselves are not necessarily the culprit, or at least they wouldn't be under different conditions, and giving people who menstruate more supposedly sanitary options doesn't help nearly as much as would simply giving back the resources we took during centuries of extraction and theft. As Bobel points out, "The bulk of MHM interventions rely too much on individual intervention through product provision and too little on structural and societal change."[28] Structures are where the power is, and if we want to support menstruating people, that's where we should put our noticing to good use.

MEDICAL MISTRUST, IMMUNE RESPONSES, AND THE SARS-COV-2 VACCINES

As I mentioned at the start of this chapter, the seesaw of paternalistic gynecology means that many people with uteruses have constantly fluctuating experiences with their care: half the time there is hand-wringing around fertility (fears of it being reduced if you are white, or having too much of it if you are not), and the other half of the time there is minimizing of their experiences as psychosomatic, overblown, or otherwise not real.[29] The noticing of gynecology, then, is largely around those ailments assumed to affect fertility. Even outside of gynecology, the structures of medicine and pharmacology do not leave room to consider that those of us with uteruses may have concerns about our bodies other than whether we can procreate. This became increasingly clear to Katie and me as we embarked on our research looking at menstrual changes post SARS-CoV-2 vaccines.

Vaccine trials tend to require that research participants not be pregnant and take measures to avoid pregnancy, like take hormonal contraceptives. The only menstrual data trials typically use to screen participants are to ask the date of the last menstrual period, again in the interests of ensuring that potential participants are not pregnant. These criteria are not bad, exactly. The problem is, from what I can tell from talking with vaccine researchers, that this is pretty much the only information collected related to the menstrual cycle. The major adverse event reporting that happens in most trials is over the first seven days. After that, if a participant experiences a symptom (like heavy menstrual bleeding, chest tenderness, or cramping) outside of the typical questions asked (since none relate to the reproductive tract or secondary sex organs), depending on the reporting method there may be no way to volunteer an unusual symptom. Add some menstrual stigma on top of this and even if a participant wants to report something unusual, they may be loath to share it with an overworked third-party teleworker trying to get through fifty more patient calls before their shift is up.

Vaccine trials still manage to capture all but the rarest of adverse events and do a great job balancing safety and efficacy. Substantial research has shown at this point that SARS-CoV-2 vaccines do not influence fertility, do not harm fetuses, do not increase the risk of miscarriage or stillbirth, and if anything improve fertility because reducing the risk and severity of COVID-19 is protective of fertility and survival.[30] But these trials will continue to miss those bodily systems about which they ask no questions. Heavy and breakthrough bleeding came as a surprise to so many menstruating and formerly menstruating people once the SARS-CoV-2 vaccines started rolling out, and it was distressing, scary, uncomfortable, and even painful. Some gender-diverse people experienced gender dysphoria. Postmenopausal people were terrified they had cancer. We need to be able to talk about this complex middle space where side effects cause distress, discomfort, pain, or dysphoria, even if they do not impact fertility.

What we are doing in our lab is providing the first sustained attention to how immune activation from vaccines might exert weird and cool (and short-lived) effects on endometrial growth, bleeding, and repair. At the time of this writing, the first wave of our survey has had over 165,000 respondents, and we've published one paper and a handful of conference abstracts.[31] Our sample is what is called a convenience sample, meaning respondents are not randomly distributed through the population, and therefore the sample could be skewed. Our survey is not a randomized controlled trial. It does not have a control group of unvaccinated people with whom we can compare menstrual cycles. In a pandemic we were not keen to encourage nonparticipation in such an important health measure, and if people want the data collected this way, they should talk to the directors of vaccine trials, who should have done this work in the first place.

A few things are clear from our sample. First, about 40 percent of cisgender, currently menstruating respondents experienced heavier menstrual bleeding after getting the vaccine; a similar proportion experienced no changes, and 20 percent reported lighter bleeding. A majority of our respondents on long-acting reversible contraception had breakthrough bleeding, as well as almost 40 percent of those on gender-affirming hormones and two-thirds of postmenopausal people. Those who were more likely to report heavier bleeding after getting vaccinated were older, had been pregnant, had given birth, or had been previously diagnosed with a major reproductive disorder like endometriosis, menorrhagia, fibroids, adenomyosis, and/or polycystic ovarian syndrome.

We found almost no differences between regularly menstruating people who were on hormonal contraceptives versus those spontaneously cycling. Between this finding and the fact that age, pregnancy, and parity increased the risk of having heavier bleeding, we suspect that the pathway by which the vaccine is affecting the uterus has more to do with temporary inflammation and coagulation changes than ovarian hormones. Essentially, anything that is associated with endometrial tissue doing more proliferating, differentiating, and repairing is going to make it so that if there is a big inflammatory event (like a lifesaving vaccine), there might just be more responsive tissue in these folks.

While my dream has long been to study these sorts of variables prospectively (rather than retrospectively, as this survey did), the funding agencies to which we have applied have not yet seen it our way. I would love to collect menstrual effluent in order to test our hypotheses about the inflammatory and coagulatory factors that probably change after some types of immune challenges such as vaccines. Perhaps a rich, feminist benefactor will help us launch that project one day. Beyond merely fulfilling my dream, collecting menses would allow us to see how the expression of some of these factors looks different locally—directly from the endometrium—than systemically—if we just take a blood sample. Blood tests from peripheral blood—that's just the blood circulating in your system—is a diagnostic standard for many conditions. You have probably had a number of blood draws in your life to check for infection, nutrient concentrations, glucose, inflammation, and more. But not all things that happen to one part of our body are necessarily going to show up in a measure that averages things across our whole body, and multiple studies have found differences between peripheral and menstrual blood.[32]

We hypothesize that some of these immune factors, especially those relevant to uterine tissue remodeling, will be different when measured in different places for postvaccine heavy bleeders. We also hypothesize that the effects of the immune response on the hemostatic system—the system that affects bleeding and coagulation—do in turn affect that major bleeding event experienced by most people with a uterus for some chunk of their lives, menstruation. These systems evolved together, and the host defense system involves coagulation because it limits how readily pathogens spread throughout the body.[33] However, it's also well established that when an immune system is overactivated, that hemostatic system also becomes overactivated. When that happens, blood clots create secondary bleeding. In light of this, we suspect something might go temporarily wonky with menstrual repair mechanisms in some individuals, especially as they relate to bleeding and clotting, when the immune system experiences an acute and powerful challenge.

Should we someday find more immune activation, inflammation, and/or lowered coagulation capabilities in heavy bleeders, and should this be found in menstrual effluent more readily than in peripheral blood, it would point to the importance of developing menstrual effluent collection protocols when patients are experiencing painful or disruptive symptoms related to their periods. It also means, as hundreds of thousands of our participants could have told those medical doctors who were quick to claim there is "no biological mechanism," that the relationship among the immune system, uterus, and ovaries is real, is relevant, and has short-term but significant effects on the lived experience of the menstrual cycle.

Part of the reason I started studying inflammation and how it affects the menstrual cycle is that inflammation doesn't just tell us about the immune system. Inflammation is a stress pathway that can be activated by many stressors, like high-intensity physical activity or psychosocial stress. In some contexts, there are benefits to stressing the body. For instance, even though bouts of high-intensity exercise stimulate inflammation in the short term, we also know that, overall, exercise promotes an anti-inflammatory profile.[34] By contrast, psychosocial stressors—to which we will turn our attention in chapter 5—do not appear to have any anti-inflammatory effects. In discussing psychosocial stress, we'll continue to consider what gets noticed and what does not. This practice will be particularly important because when it comes to psychosocial stressors, menstruating people are even more likely to be told that they are imagining something that isn't there.

CHAPTER 5

Big Little Stressors

DURING THE Senate confirmation hearings for Supreme Court justice nominee Clarence Thomas, Anita Hill, his former assistant and herself a lawyer, testified that Thomas had sexually harassed her while working as her supervisor at the Department of Education and the Equal Employment Opportunity Commission.[1] The study of sexual harassment was relatively new at the time, and one common but misguided belief was that a failure to quit, or a failure to cut off one's relationship with the perpetrator, provided evidence that there had been no harassment. Senator Alan Simpson (R-Wyoming), for instance, pressed Hill about why she kept working for Thomas, in an attempt to discredit her testimony:

> SIMPSON: But let me tell you, if what you say this man said to you occurred, why in God's name, when he left his position of power or status or authority over you, and you left in 1983, why in God's name would you ever speak to a man like that the rest of your life?
>
> HILL: That is a very good question, and I am sure that I cannot answer that to your satisfaction. That is one of the things that I have tried to do today. I have suggested that I was afraid of retaliation, I was afraid of damage to my professional life, and I believe that you have to understand that this response—and that is one of the things that I have come to understand about harassment—that this response, this kind of response, is not

atypical, and I can't explain. It takes an expert in psychology to explain how that can happen, but it can happen, because it happened to me.[2]

As the exchange makes clear, Simpson, like many people then and now, did not understand typical behavioral responses to harassment nor that Hill's response was a typical one. Most people don't quit their jobs; they might minimize and endure and even maintain relationships with their perpetrator.

Dr. Lilia Cortina was a psychology major at Pomona College during the hearings and decided to pursue graduate school to study sexual harassment as a result of listening to them. Now a professor at the University of Michigan, she has published extensively on the topic. In one piece, Cortina and her coauthor Dr. Vicki Magley explain that the main forms of resistance available to sexually harassed people are "exit," or leaving the organization, and "voice," or expressing resistance.[3] One offers a permanent solution but no paycheck—the other risks significant retaliation. This means that the most safe and common form of coping for many people dealing with sexual harassment is not resistance but avoidance. In another paper, this time on rude behaviors of ambiguous intent, Cortina and Magley found a variety of behaviors by targeted people that were avoidant of the perpetrator: people seek support from others, they detach emotionally from the environment where harassment is happening, they minimize the extent of the harm, or they find ways to avoid further conflict.[4]

Lilia has been driven in her work to explore two major aspects of harassment and other negative workplace behaviors: that people who are harassed rarely respond openly or loudly, and that the harms they experience are real. Rudeness, in particular, is often seen as trivial—most of all by those who hold power or are adjacent to it within an organization, but in the past this misperception of triviality has been an issue among researchers who study counterproductive workplace behaviors too. Rudeness could look like a coworker leaving you off the social email to go out for drinks, downplaying your contributions in a presentation, or maybe pronouncing your name or title incorrectly despite multiple corrections. Because of the ways rudeness is theorized as a minor transgression within the social sciences, researchers only discuss it under circumstances where it might eventually trigger an equivalent or escalated response, as defined by those researchers. Historically in social sciences like psychology and anthropology, research has focused on aggressive behavior rather than, say, the avoidant behavior Lilia and colleagues were finding to be the more common response to rudeness and harassment.

Few even seemed to recognize that the harms from rude behaviors were differentially experienced. As Lilia puts it, "This general incivility wasn't entirely general. People interpret it as identity neutral because there was no sexism or racism or anti-gay rhetoric." But these seemingly ambivalent behaviors are as much about identity as sexual, racial, or LGBTQ+ harassment is about identity. Lilia, along with other researchers, started noticing the gendered aspect of rudeness during a study among federal court attorneys. "The women attorneys were describing both far more frequent sorts of interruptions and being ignored and being called by their first name when their co-counsel gets called Mister So-and-so." How did the men describe experiences of rudeness? "More like, 'the judge didn't accept their sound legal argument,'" said Lilia with a chuckle.

Triviality is a core construct of the paternalistic way many systems approach women and gender-diverse people. That is, if we can't make this issue about fertility but it has to do with gender/sex, the issue gets downplayed, made insignificant, or treated as frivolous. Trivializing the experiences of women and gender minorities is common even when people are in excruciating pain. In her book *Ask Me about My Uterus: A Quest to Make Doctors Believe in Women's Pain*, Abby Norman points out that there is a long history to women struggling to have their experiences taken seriously. "If women have become synonymous with hysteria, malingering, and hypochondria in a clinical setting," she writes, "then it has far less to do with the natural inclinations of women and behavior than it does with the history of medicine."[5] In emergency rooms, doctors are slower to administer pain medication to women compared to men and slower to administer pain medication to nonwhite patients.

When people express a desire to prematurely have their IUDs removed, doctors often cajole them into keeping them in or passively (and sometimes directly) express their disappointment at their lack of fortitude.[6] And while there has been an established link between hormonal contraception and depression for several decades, the second that a similar link was established in clinical trials for "male" hormonal contraception, researchers halted the study.[7] I find myself "noticing" how systemic invalidations of lived experiences fall along race and gender lines. This leads me to two melodies worth drawing out: that both the stressors we experience and the way they are perceived within the systems where we seek care have repercussions for our bodies. This plays out in key ways with menstrual cycles.

Psychosocial stressors—such as those menstruating people experience at work and the way they experience their bodies in relation to medicine—can have acute and lasting effects on our systemic inflammation, our hypothalamic-pituitary-adrenal (HPA) axis, and the various set points at which our body functions, disrupting flexible adaptations we've developed to handle our environment and appropriately allocate resources throughout the body. As communal animals, our response to these stressors is an evolved trait that is deeply social, related to our positionality within our communities, and affected by how and whether our grievances are heard so that we may develop appropriate coping mechanisms to heal. This stress response and how we are situated have particular implications for the ovaries, uterus, and downstream target tissues because of how survival and belonging are tied up in reproduction. Therefore, fundamental to understanding how psychosocial stressors affect the menstrual cycle is addressing the ways in which women and gender-diverse people are so often not believed. The work we must do to hide our pain, distress, or trauma to avoid being called weak; strongly state our pain to receive appropriate treatment; endure our pain because of how long we must wait to receive care; alternate between these states as we brilliantly recognize context and take care of ourselves as best we can? Anyone with chronic illness recognizes this as the work of a lifetime.

In this chapter I characterize psychosocial stressors and demonstrate the effects they have on the material body, most of all the menstrual cycle. As with the discussions I raised in chapter 4, the effects of stressors on cycles are complex. Psychosocial stressors do not typically exert some sort of one-to-one effect on menstrual cycles. Instead, for example, medical abuse in childhood or high-frequency sexual harassment leads to a chronic stress profile that affects how the body reacts to current stressors, often making one more physiologically reactive. How this manifests is highly variable across people and populations. The important thing is to do the work of noticing psychosocial stressors and the structural oppressions that often produce them. Both gender/sex and racial health disparities were assumed for a long time to have a biological origin, but we now know that these disparities are social in origin, and the mechanism through which they act relates to psychosocial stress.[8] As epidemiologist Dr. Whitney Robinson put it to me recently, "It's not genetic inferiority. It's because of things that happen to people. That means we do not have to accept it as an inexorable deficit of health; there is something to be changed."

"SUBTLE" IS NOT "TRIVIAL"

Compared to energetic or immunological stressors, psychosocial stressors can be harder to define and conceptualize. The ways we have conflated subtle and trivial are part of the problem in how we understand the effects of stress on the body. Some psychosocial stressors are trivial, in the sense that they are transient: for most, studying for a big test is stressful but leaves no lasting harm. One study assessed the anxiety and stress scores, as well as cortisol, estradiol, and progesterone concentrations, of women studying for and then taking the MCAT (the medical school entrance exam in the United States) to determine the extent to which this psychosocial stressor influences the menstrual cycle.[9] No matter the stress measure—anxiety, stress scores, or cortisol—nothing seemed to affect follicular estradiol or luteal progesterone in this sample (you may remember the follicular phase is the first half of the cycle, from menses to ovulation, and the luteal phase is the second half of the cycle, from ovulation to the next menses). In this case, the stress and anxiety the students were experiencing were not manifesting physiologically in significant changes in cortisol and therefore changes in HPA activity.

However, there are subtle stressors out there that are far from trivial. Scientists have devoted much of our time to characterizing the body's responses to seemingly more severe events, so it follows that we have struggled to understand more subtle stressors. Early research on psychosocial stressors primarily paid attention to life-threatening events—and, in particular, life-threatening events that would have been common when our species was young: predation, physical fights over mates or territory, the physical stress of hunting, and maybe periodic starvation from hunger seasons. This focus on what we might think of as the Big Bad Stressors limited our understanding of psychosocial stress and created the perception that a psychosocial stressor could only be meaningful if it represented an immediate threat to survival or reproduction.

On the one hand, it's true that Big Bad Stressors create a Big Bad Response in us: they can trigger fight (take on the stressor), flight (run away from it), or freeze (far more common than we realize).[10] These behavioral responses are all supported by activation of the HPA axis via cortisol, as well as the sympathetic nervous system via epinephrine and norepinephrine. We stop certain processes like digestion so that we can shunt all that energy to our lungs to breathe or to the muscles in our legs to run. You probably know this sensation—for me it is an uncomfortable warmth that starts in the center of my chest and spreads, and I can feel my heart pounding.

In addition to studying Big Bad Stressors, early anthropologists paid a lot more attention to male animals (including humans) than anyone else. They paid attention to the most obvious, most violent moments: for instance, how chimpanzees rip each other's balls off during territorial defense (really), or how baboons compete for mates, or how some orangutan males sexually coerce females to increase reproductive opportunities.[11] Male primates were seen as more gregarious, more interesting, and their behaviors more relevant to our understanding of human evolution than female primates.[12] Eventually, in large part thanks to feminist primatologists, more scientists recognized the importance of the smaller moments for all genders of primate: the shuffling to turn your back on someone, the denial of food from another begging, the huddling with kin to soothe and comfort, and other everyday moments that determine relationships, rank, and quality of life.[13] These are moments primatologists characterize as appeasement, reconciliation, agonism, even befriending, and as in humans, these everyday behaviors and responses are the social glue that hold together primate societies.

These are the moments that Drs. Lilia Cortina, Sandy Hershcovis, and I engaged with in a 2021 theoretical paper.[14] Imagine someone is rude to you at work, in a way that really hurts your feelings. This interaction doesn't just hurt your feelings, however. It also activates your stress response. What you do next will determine whether that stress response is amplified or cut short. Increasingly, it looks like hostile or avoidant responses will keep you on that path of stress, while choosing an affiliative response is more likely to calm your systems down.[15] By an *affiliative response*, I mean many things: approaching the person who hurt you to talk it out, venting to a trusted friend, or problem-solving with a mentor. Of course, how often these options are available to you will depend upon the culture and makeup of your workplace. A workplace with what is called a *masculinity contest culture* incentivizes workaholism, fosters overcompetition, prizes physical strength and stamina, and punishes weakness. These workplaces are unlikely to leave room for kindness or talking out hurts. As another example, a woman of color in an otherwise all-white workplace may struggle to find someone who can offer social support.[16] It becomes a question of access: in certain workplaces it's not safe to reconcile, only escalate; it's not possible to find someone with whom to vent because no one will believe you. For a person who experiences marginalization in a given workplace and is therefore more susceptible to being targeted, this is a double whammy: they experience more harm and have fewer resources available to reduce or repair it. This is why subtle is not trivial.

Finally, the way historically excluded groups have been acculturated to respond to rude behaviors may further limit their ability to take action: people who are discriminated against, because of a lifetime of being gaslit about these harms, are more likely to attribute the experience to their own personal failings.[17] Rather than feel anger, which often calls people to action, people who are discriminated against are more likely to feel shame. In one experiment, college-aged women were put into a situation where they were guaranteed to fail at a task. Conditions were manipulated so that for half of the women it was clear their failure was related to gender discrimination, thus an external source and not their fault. Shame, not anger, was exacerbated among the women for whom discrimination caused their failure.[18]

In a follow-up study to this experiment, when researchers primed feelings of shame in women, they responded to the task failure with increased cortisol reactivity (meaning, a stronger cortisol response to the stressor) and were less able to act. If the systems under which we live and work increase one's risk of being targeted, limit the ability to find social support, and teach us we are wrong to externalize mistreatment, then those subtle harms are part of the broader pattern of systemic harm that many people experience under colonialism and capitalism. This suggests that those stressors associated with racist, sexist, cisnormative, heteronormative, ableist, and other oppressive behaviors are more likely to have echoing and lasting effects via HPA and inflammatory systems on many parts of the body. Fundamentally, injustice is out-grouping that denies many people the community they deserve and need for survival.

Not being able to affiliate after a stressful event has many physiological and mental health consequences. Bullying and discrimination can lead to anxiety, depression, withdrawal from others, and burnout; as just mentioned, they can even affect our cortisol reactivity. Many associational studies have found that measures of psychosocial stress, past and present, are higher among people with menstrual pain, irregular cycles, and even amenorrhea (recall this means the total loss of the period).[19]

A more recent experimental paper that makes it possible to test hypotheses about cause and effect looked at social exclusion and its possible effects on menstrual cycle hormones.[20] A few days after a battery of social surveys that included an assessment of their current social support, researchers put participants (UCLA students) into a group setting where they answered increasingly personal questions as a way of rapidly developing bonds among them. Then, each participant was sent back to an individual computer to rate the people they had met. The devious part (I was surprised this made it past their office of human subject protections) was that they sorted participants into social inclusion/social exclusion experimental groups, and the ratings they received from their peers were all faked. Participants in the social inclusion group received comments such as "I liked her" and "She was friendly," while those in the social exclusion group received comments like "She seemed full of herself and annoying," and "I don't know how she got into UCLA." According to the authors of the study, "These reasons were pre-rated by female undergraduate research assistants for realism and amount of stress evoked." Personally, I felt my cortisol—and sense of shame—rise just reading them.

The participants collected a baseline saliva sample and then another after reading these nice or cruel statements about them. Participants who reported having less social support in their daily life had changes to their estrogen to progesterone ratio depending on whether they had been manipulated to feel socially included or excluded. In the social exclusion group these participants displayed a drop in their estrogen to progesterone ratio, while those in the social inclusion group displayed an increase in their estrogen-to-progesterone ratio. This ratio change was mostly explained by changes in progesterone concentrations, which matches other evidence that adrenally excreted progesterone (those same adrenals of the HPA axis) might rise with stress. Follicular progesterone (often assumed to be from the adrenals) has been shown to be elevated among those with early pregnancy loss.[21] These effects also seemed to be concentrated among individuals who happened to have their study day occur around midcycle. Other research suggests that midcycle is when menstruating people are more aware of positive and negative feedback (counter to what a lot of people assume, which is that the premenstrual phase is when we are "sensitive").[22]

Another study looked at cortisol concentrations and subjective stress measures, again comparing follicular to luteal phase responses. Participants were subjected to a modified Trier Social Stress Test—in this case, a mock job interview in front of two evaluators who were trained to provide neutral expressions. From ten minutes before the interview through sixty minutes afterward, at ten-minute intervals, participants filled out a subjective stress survey and provided a saliva sample for cortisol. Those who experienced the mock interview in their follicular phase exhibited a negative correlation between their peak cortisol and subjective stress measures, meaning those with the highest cortisol actually self-reported the lowest stress scores. Those who were tested in the luteal phase had the opposite relationship between cortisol and subjective stress: people with the highest cortisol tended to have the highest self-reported measures of stress.[23] The messaging I have received about the premenstrual phase (the last week of the luteal phase) for most of my life is that it is a period of irrationality and unreasonable anger, which I always took to mean that premenopausal people have outsized responses to small insults. It's interesting to me, then, that this research suggests people's objective and subjective measures of stress are at their highest concordance at this time in the menstrual cycle. If what is happening internally in terms of our HPA stress response and how we experience it subjectively are not in opposition in the luteal phase, where does that leave the premenopausal stereotype?

To summarize: psychosocial stressors have repeatedly been shown to correspond to menstrual cycle changes and negative symptoms. Experimental and observational research that allows us to look at this relationship more causally suggests that the timing and type of stressor matters to whether it affects the menstrual cycle. But the other big conclusion here is that the premenstrual phase has not been found to be a time when people (at least cisgender, young university students in the United States) are more sensitive or reactionary to those stressors. The evidence that the luteal phase, especially the late luteal or premenstrual phase, is not necessarily a time of increased sensitivity makes me wonder if the way psychosocial stressors affect our cycles has less to do with whether it affects gonadal hormones and more to do with how it affects our experience of our cycle itself.

ARE YOU PREMENSTRUAL OR SOMETHING?

In *Dear Science and Other Stories*, Dr. Katherine McKittrick writes of how "in many academic worlds categories are organizational tools."[24] These categories create disciplinary thought, and McKittrick explains that this very structure is self-reinforcing; that is, we create categories so that we can design studies and ways of thought that then serve to support those categories. For instance, when we create a conceptual group like "race" that is produced in a system that believes in biological determinism and was created under colonialism, we can unwittingly support biologically deterministic and colonialist ideas about "race" instead of revealing the overarching and causal structure of racism. When it comes to the menstrual cycle, we have a category, "premenstrual," that has no boundaries related to the functioning of the ovaries or uterus—it does not, for example, map to anything like "follicular" or "luteal" phase, which bound the timing of the folliculogenesis and corpus luteum appearance, respectively. *Premenstrual* as a concept has to do with how we experience our cycles or how others assume we experience them.

The boundaries of the category "premenstrual" tend to be temporal: the seven days or so before menses (though more on that in a moment). If we are to ascribe any biological feature of this time, it is that in cycles that are both ovulatory and nonconceptive, progesterone is declining in concentration because the corpus luteum, which supplies most of our progesterone, is breaking down without the presence of an embryo to ask it to stick around. Once it's gone, so is the main source of progesterone, which is a large part of what stimulates the beginning of endometrial tissue breakdown and the healing processes, leading to the menstrual period.

The idea that menstrual cycles universally involve high midluteal-phase progesterone followed by a terrible crash at menses initially led to the hypothesis that the cause of premenstrual syndrome (PMS), premenstrual dysphoric disorder (PMDD), and even postpartum depression (PPD; because the placenta is an even bigger source of progesterone than the corpus luteum) is related to the loss of progesterone in each of these time periods. PMS is defined as recurrent, moderate-affective, physical, and/or behavioral symptoms that resolve quickly upon menstruation; PMDD as recurrent, moderate-to-severe affective, physical, and/or behavioral symptoms that affect daily functioning and for some linger into menstruation; and PPD as depression that appears in the peripartum period (about a third of those who develop PPD begin to experience symptoms before giving birth).[25] Yet PPD, PMDD, and

PMS are not universal: cross-cultural estimates range from 0 to 60 percent for PPD, 13 to 18 percent for PMDD, and 10 to 90 percent for PMS, where the highest percentages correspond to individuals reporting just one symptom.[26] PMDD and PMS also suffer from retrospective analysis bias; that is, most studies use methods that inquire about respondents' symptoms and severity in prior menstrual cycles. Compared to prospective analysis, in which people track their symptoms and severity on a daily basis, retrospective surveys tend to overestimate the amount of PMDD and/or PMS in a given population.[27]

Defining and diagnosing PMDD and PMS demonstrate what happens when you take a fairly common experience and wedge it into a dominant science construct. On the one hand, there are many reasons to be cautious when putting someone into these categories, and to create a stringent diagnostic definition for PMDD in particular, in order to avoid pathologizing menstruating people. On the other, telling some people their self-described experience is not real denies their lived experience and their own characterization of their bodies.

Dr. Jane Ussher, a psychology professor at the University of Western Sydney, Australia, has taken on the problem of quantitative measurement of PMDD. She writes that we need to "unravel the process by which women come to understand the reproductive body as a cause of disorder or distress" and ultimately how we come to characterize ourselves as belonging to certain (often subjugated) categories.[28] Ussher found that the strict criteria that create the PMDD category lead to categorizing some sufferers as "false positives" or "hypervigilant." This approach puts all of the control in the hands of practitioners and researchers and none in the hands of patients and research participants. In interviews meant to capture more nuance in how people narrate their experience of premenstrual symptoms, many participants replicate the list of symptoms found in any PMDD checklist. But Ussher also found that participants begin to disclose that the premenstrual phase is a time of "problems in relationships, problems at work, needing time to themselves, feelings of overwhelming responsibility, and failing to cope and be in control at all times." Ussher and others have also shown that there are differences in how people with premenstrual symptoms contextualize their experiences based on their intimate relationships. While partnered heterosexual women with premenstrual symptoms, especially those with children, can report feeling overwhelmed by family demands and their symptoms doubted by their husbands, partnered lesbian women with premenstrual symptoms are more likely to report experiencing nonjudgmental acceptance and explicit attempts to relieve distress from the same symptoms.[29] What seemed to be an important distinction between the samples was how the partners recognized and validated the symptomology. Being trusted and believed are powerful experiences, which affect how one experiences negative symptoms.

Context, too, matters in other ways to the experience of the premenstrual phase. In one sample, researchers found that US ethnic minority women who were born outside the United States had a lower risk of PMDD compared to US-born participants. In fact, the longer participants lived in the United States, the higher their risk for PMDD.[30] Another analysis by this research team showed that those participants who reported that they had experienced discrimination (due to race, gender, or other factors) were more likely to have PMDD and more likely to have premenstrual symptoms.[31] In fact, subtle discrimination was more strongly associated with PMDD than blatant discrimination. According to these two studies, being taught negative attitudes about menstruation and experiencing discrimination, both prevalent in the United States, can increase the chance that one experiences premenstrual symptoms and possibly PMDD.

Psychosocial stressors affect the menstrual cycle not just in the ways that energetic and immune stressors do (via resource allocation) but by moderating our experience of the cycle itself. Dr. Tory Eisenlohr-Moul, a psychiatry professor at the University of Illinois at Chicago, studies those 6 percent of menstruating people who with prospective quantitative analysis can be diagnosed as having PMDD. It is incredibly important to distinguish between premenstrual symptoms, which many people report experiencing (and should be acknowledged as experiencing), versus what Eisenlohr-Moul considers a rare set of people whose brains struggle to adapt to the typical hormone fluctuations of the menstrual cycle. And yet, as she points out, "It's not one thing." There are a range of psychological, somatic, behavioral, and emotional symptoms, even among the smaller number of prospectively diagnosed PMDD sufferers. Eisenlohr-Moul and colleagues have also found that the timing of PMDD is a bit squishier than we've presumed in the past: while the majority (65 percent) of people who experience PMDD have moderate symptoms the week before their period, 17.5 percent have two full weeks of severe symptoms, and another 17.5 percent have one week of severe symptoms that don't resolve until a few days into menses.[32]

Eisenlohr-Moul and colleagues suggest that the existence of these subtypes might be why there have been such mixed results in different drug trials to treat PMDD, such as hormonal contraceptives to regulate hormone fluctuations or selective serotonin reuptake inhibitors, a class of antidepressant. The existence of different subtypes points to hypotheses for the origin of PMDD that go beyond sensitivity to progesterone withdrawal: one subtype of PMDD may develop a special sensitivity to hormones more broadly, and another subtype may be especially affected by allopregnanolone, a neurosteroid that progesterone converts into during the latter half of the cycle.[33] As Eisenlohr-Moul notes: "This makes the conversation about whether this is genetic, or environmental, or both, way more complicated, because which subtype are you talking about?" It's helpful for people who are suffering with suicidal ideation, uncontrollable mood swings, and more to receive an official diagnosis. Yet as is the case with so many other syndromes that relate to the ovaries or uterus, what we seem to be doing is putting together similar-looking conditions with different backstories into a single category.

One other factor that contributes to all this variability in how PMDD looks across people is related to their histories of abuse, not only because of how early and/or severe trauma can affect the stress response but also because it can indicate one's familial environment. Eisenlohr-Moul reminds me that something that can co-occur with early abuse is that the abused person may not have had anyone to teach them good emotional-regulation strategies. Eisenlohr-Moul and colleagues have found that physical abuse and sexual abuse interact differently with estradiol and progesterone to affect mood over the menstrual cycle among people with a menstrually related mood disorder.[34] Those who had

experienced physical abuse had greater mood and interpersonal symptoms that corresponded with the luteal increase in progesterone. Those
who had experienced sexual abuse had greater anxiety symptoms
around the midcycle rise in estradiol, which for many is when they experience greater sexual awareness and desire. As Eisenlohr-Moul points
out, "If you have a history of sexual trauma, feeling increased sexual
desire can be complicated." Not everyone particularly notices their pattern of ovarian hormone shifts throughout the menstrual cycle, and
most of those who do experience only a few mild changes. But for a
small number of menstruating people, their trauma histories combined
with their individual biology can make these pattern shifts disruptive,
can strain relationships, and can even lead to suicidal ideation.

TELL ME ABOUT YOUR CHILDHOOD

We don't know why some people may be more susceptible to PMDD
than others, but the hormone sensitivity hypothesis is reminiscent of a
similar hypothesis about stress reactivity and how it may affect puberty.
You may recall in chapter 2 that I discussed how the timing of the first
menstrual period has been shown to vary under different environmental and contextual conditions; that is, in environments where father
absence may be a psychosocial stressor (for instance, where father absence is a result of the carceral state and thus can cause significant familial trauma), it may drive menarche early, but in environments where
father absence is not an adverse experience (for instance, in cultures
with care provided by many outside the nuclear family), it has no real
effect. Some early evidence from our lab even suggests that strong
parent-adolescent attachment can swing menarche early or late, depending on the parent.

As I've mentioned, menarcheal age is just one part of a massive, ongoing process of puberty that starts with adrenarche and does not conclude until your bones fuse or until you reach certain socioemotional
benchmarks. There are many different factors that grow or change during this time that could be used to measure pubertal timing (the age at
which the body starts certain processes/hits certain benchmarks) and

tempo (how quickly the body moves through different processes). Pubertal timing could be measured by when the first pubic hairs appear; pubertal tempo could be measured by how quickly the body goes from stage one to stage five of pubic hair growth. The effects of childhood stressors on puberty have been extensively studied because puberty seems to be so malleable. Puberty timing and tempo are still plenty heritable, but it is also going to be affected by what's happening around you.

In addition to familial and cultural contexts, a person's particular reactivity to stress also modifies the relationship between the stressor itself and puberty. Stress reactivity is defined as how one responds to a stressor: how they tend to respond and the extent to which a stressor invokes a threat response. Stress reactivity varies with a person's individual biology, their childhood environment, and the extent to which they had access to learning emotional regulation strategies from trusted adults. Stress reactivity is often lower among people who have experienced early adversity or trauma.[35] Stress reactivity can be measured in a number of different ways, but the most common is by looking at cortisol reactivity: expose a person to an experimental stressor and examine how quickly their cortisol elevates, by how much, and for how long. People who are more stress reactive may have a stronger, faster, and/or longer cortisol response.

Changes in stress reactivity seem to depend on the specific stressor and frequency of exposure. In one study of male children of alcoholics (participants ranged from eight to sixteen years old) in India, cortisol reactivity was blunted, meaning the initial cortisol response was not as high as age-matched controls.[36] In another study, using the Whitehall II cohort of British civil servants (the average participant age by the time the study concluded was sixty-three years), researchers explored cortisol response among people with and without early adverse experiences. Among those participants with early adverse experiences, the researchers were also able to study those who had experienced no, few, and many bouts of psychological distress throughout their adult lives.[37] Those with early adverse experiences who also had bouts of psychological distress in adulthood had blunted cortisol reactivity. Those with early adverse experiences who had not experienced significant psychological distress in adulthood did not have that blunted reactivity, but they did have higher baseline cortisol and more prolonged cortisol responses to stress.

Dr. Bruce Ellis, a professor of family studies and human development at the University of Arizona, and his colleagues have shown that stress reactivity affects the relationship between the quality of parent-child relationships and both pubertal tempo and timing.[38] Because early adversity tends to be associated with earlier pubertal timing, one would expect a more secure environment to be associated with later timing. While past studies of parent-child relationships and pubertal maturation found associations between the two, these associations had small effect sizes, indicating the correlation is not that strong.[39] One possibility is that individual differences between children are sufficiently strong to affect this finding. To assess this possibility, Ellis and colleagues hypothesized that those children with greater stress reactivity would have stronger pubertal responses to the quality of the parent-child relationship. They found that among those with higher stress reactivity, those with better parental relationships had a slower pubertal tempo and later pubertal timing. Those with lower stress reactivity and good parental relationships had the opposite pattern: a faster pubertal tempo and earlier pubertal timing. Stress reactivity is not necessarily "good" or "bad," but it can tell us how responsive someone may be to different environments. And again, it's quite heritable, meaning that to some extent you get what you get: if you happen to be someone with blunted stress reactivity, it does not necessarily follow that you experienced substantial early trauma.

Over the past few decades, one of the main ways people have tried to understand the influence of early psychosocial stressors on adult health is via the Adverse Childhood Experiences questionnaire, or ACE.[40] The ACE captures several types of early adversity and trauma: sexual abuse, physical abuse, emotional or physical neglect, emotional or physical abuse of one's mother, alcoholism or drug use in the home, loss of a biological parent due to divorce or abandonment, mental illness of a parent, and the incarceration of a close family member. Each time a participant answers yes to the presence of one of these experiences, their ACE score goes up by one. An ACE score of zero or one tends not to be associated with any major health risks. But a number of studies have shown that those with an ACE score at or higher than two are at increased risk of heart disease, depression and anxiety, some cancers, and diabetes, with other studies suggesting a threshold of four ACEs.[41]

The main issue with the ACE is that it mixes many different types of stressors and treats them equally. However, we know that not all adverse experiences have the same impact, so merely looking at the additive score can be misleading. For instance, childhood maltreatment (for example, abuse or neglect) often has a stronger effect on health than household dysfunction (for example, divorce, incarceration, substance use).[42] More generally, a recent systematic review and meta-analysis of the cortisol reactivity and adversity literature suggests that there is an important difference between the effect of adverse and traumatic events.[43] Adverse events, like divorce or moving homes, may be more likely to lead to a hyperreactive cortisol response. Traumatic events, like sexual or physical abuse, may be more likely to lead to a blunted cortisol response. Our bodies learn to respond to our environment when we are young with the expectation that this environment will stay the same as we age. If we are very physically active and/or experience many immune challenges or psychosocial stressors, our young bodies interpret this as the typical environment for us. We adapt to these early experiences and modulate hormones, receptors, gene expression, and nervous and inflammatory pathways so that we are adequately prepared to encounter them again. In many cases, this is a great feature of our growing bodies, but the relationship between early adverse experiences and later health suggests that sometimes these functions become dysregulated. If we are exposed to too many stressors, our bodies can learn to anticipate or expect terrible things all the time, develop a blunted response to stress, or slow our metabolism and retain energy in anticipation of later famine. As a result, we can develop constantly activated inflammatory pathways and activated or blunted HPA responses.

In my lab, we administered a modified version of the ACE (reflecting certain cultural differences) to the rural Polish women with whom we work. In our preliminary analyses, an earlier age at menarche was associated with a higher ACE score, but only if the model also considered other factors that might influence inflammation, such as whether participants had performed farmwork as children, an experience that should carry significant immune challenges from outdoor and soil microbe exposure, as well as energetic demands from the increased physical workload.[44] Several different types of stressors—energetic, immunological, and psychosocial—together were needed to exert an effect on age at menarche. This general idea—that stressors must be studied alongside one another to understand their effect—has been supported by a range of ACE research. One single adverse childhood experience, while of course meaningful and worth paying attention to, does not necessarily leave significant traces on the body.

The effects of adverse experiences on adult menstrual cycles are less strong than on age at menarche, but they are still observable. In our rural Polish sample, we found that women who reported three or more ACEs had significantly lower estradiol compared to those with no ACEs.[45] Some of our attempts to understand the relationship between early life events and adult estradiol have also shown that ACEs pop up in some of our models, so long as, again, certain farming variables that we know affect similar stress pathways are also included. Other researchers have found that women with more ACEs tend to self-report having had more irregular menstrual cycles in their past.[46]

Early life experiences may also help us understand physical pain associated with the uterus. Chronic pelvic pain is defined as pain experienced in the pelvic area that lasts for at least six months. Chronic pelvic pain is also often categorized as dysmenorrhea (pain during menstruation), dyspareunia (pain during penetrative intercourse), and noncyclical pain (of a more general nature). While dysmenorrhea is the most common, affecting over 90 percent of subjects in some samples, dyspareunia and noncyclical pain present prominently across many samples as well.[47] Women with chronic pelvic pain in their reproductive years are more likely to have previously experienced sexual abuse.[48] Those with two or more ACEs also tend to experience more somatic pain generally.[49] One of the common eventual diagnoses (because it can take so many years to be formally diagnosed) underlying chronic pelvic pain is endometriosis, a condition in which lesions of endometrial tissue live outside the uterine cavity but respond to ovarian hormones, often causing pain, lower gastrointestinal issues, and/or infertility. Two different studies have shown that there is no particular relationship between the number of endometriosis lesions in a person's abdominal cavity and the severity of their pelvic pain.[50] Some people suffer greatly with a small number of lesions, while others have endometriosis everywhere in their abdomen but do not experience much, if any, pain.

This suggests that at least some portion of the pain experienced in both chronic pelvic pain, generally, and endometriosis, specifically, may have some underlying origins in early adverse experiences. A recent study supports this conclusion. In an analysis of the Nurses' Health Study II, a data set of over sixty thousand premenopausal women, scientists found that those with severe physical or sexual abuse in childhood or adolescence were significantly more likely to be diagnosed in adulthood with endometriosis compared to those without abuse histories.[51] Among those who went to their doctors for a diagnosis because they were experiencing pain, the relationship between abuse and endometriosis was even stronger.

Throughout this section, I've been talking about these early adverse experiences as though they themselves cause menstrual cycle irregularities or pelvic pain. But it's important to recognize these as population-level correlations that are, in fact, highly unlikely to have a direct causal pathway. The idea that our early experiences modulate our hormone sensitivity and/or stress reactivity is a compelling, mechanistic way to think through how our lived experiences affect our downstream responses to our environment: like my earlier work with Lilia and Sandy, life is not just about the initial insult but what tools you have to respond to it.

Stress reactivity—and maybe some subtypes of hormone sensitivity—may be related to the ways young bodies respond to and make sense of early adversity. An anthropologist like me may look at these data and see life history strategies to make sense of and respond appropriately to one's environment; a clinician may see the beginnings of pathology; a trauma-informed therapist may see adaptations that served the child but may not serve the adult. In all cases it's important not to draw causal lines between early experiences, cognitive or somatic responses, and adult functioning whereby all people with early trauma are guaranteed to have blunted stress reactivity and later psychological challenges.

Beyond the general difficulties of applying a population-level finding to an individual, ACE research has also encountered more specific criticism. One recent paper criticizes the ACE as a chaotic concept, an abstraction that "conflate[s] different issues, or divide[s] up indivisible processes, leading to problems in their explanatory weight and hence in developing policy and interventions on their basis."[52] As the authors of the paper point out, the scales encompass a wide variety of experiences, and the yes/no nature of the questions can flatten experience. For example, if someone reports that their parents were ever separated or divorced, it "constitutes an ACE whether it was amicable or adversarial, or occurred before the respondent was born, when a toddler, or a teenager." As with PMDD, there are also problems of retrospective versus prospective study design. And because it is so tempting to see ACEs as the source of later health issues, policy is increasingly being developed to see these ACEs as a problem of epidemic proportions that must be stopped, without an intersectional and systemic appreciation of the racial, socioeconomic, gendered, colonial, and other forces that lead to some people and some communities experiencing more (and more severe forms of) ACEs than others. Seeing ACEs as the origin story for later harms provides fuel for an interventionist state response that hastily removes children from their families in anticipation of their experiencing an ACE without consideration of how family separation may create rather than reduce harm.

Another collaborator of mine, Dr. Ruby Mendenhall, professor of sociology and dean of Diversity and Democratization of Health Innovation at the Carle Illinois School of Medicine, began her career as an occupational therapist in the pediatrics department at Cook County Hospital. As part of their protective services team, Ruby witnessed many children coming into the hospital being diagnosed with failure to thrive. "And me and the medical system would say, 'what's wrong? Do you not know how to take care of your baby?'" she recounted. "We didn't say it in those words, but that's the way it came across." The mothers would explain, over and over, that they could not afford formula and that they had to water it down to make it last for their babies. The longer she worked there, the clearer it became that "it isn't about the mothers, they are capable. This is a societal issue. When will we pay these women enough money to feed their children?" These experiences were what motivated Ruby to move into policy work and eventually into sociology.

With period poverty, with children's health, and with so much more, public policy so often moves toward an interventionist response that policy-makers hope addresses a single symptom. But nearly all of these symptoms are symptoms deriving from poverty, racial oppression, and gender oppression. If we want to address the stressors that harm menstrual health—be they energetic, immune, or psychosocial—we should focus on reducing poverty and the structures that reinforce racism and sexism. As Ruby points out in her own work, in the United States we spend far more money addressing the illnesses caused by poverty than by addressing poverty itself.[53]

FAT STIGMA IS MORE HARMFUL THAN FAT

The importance of the effects of psychosocial harms on the menstrual cycle can clearly be seen in the social stigma attached to fatness, itself part of the story of what got us to the ACE instrument. Dr. Vincent Felitti was a medical doctor who saw obese patients with health concerns and tried to help them lose weight as part of their treatment plan. However, he had little ability to predict who would respond well to weight loss treatments. Some were able to lose weight over time and keep it off, while others struggled to lose weight or regained it quickly. According to Felitti, one of his patients lost an enormous amount of weight—she went from over 400 pounds down to 132—but within a few weeks quickly began regaining the weight. When Felitti asked this patient why she was regaining weight, she eventually revealed that she had been sexually propositioned by an older man at work. Felitti was curious as to why this event would stimulate over thirty pounds of weight gain in just a few weeks. Over time, it emerged that this patient had a sexual trauma history: she had been molested by her grandfather for many years as a child and young adult.[54] This was not the only patient for whom Felitti began to understand that weight was tied to early trauma. He began to find that many of his patients, more than half, had a history of sexual trauma.[55] Many of them developed coping mechanisms to deal with their trauma that involved eating. It was because of this that Felitti began to see early trauma as a precursor of later disease (which to him included being of a higher weight) and why the ACE instrument was developed.

Not all people who have experienced trauma have particular health issues in adulthood, and not all people with health issues in adulthood have experienced trauma. I say this because it has become very tempting, in our fat-phobic society, to see all fat people as in some way damaged or dysfunctional. It's an easy shorthand that only causes more harm: cruelty from a fat-phobic society and attempts to lose weight both have profound physical costs. Substantial research at this point has shown that the harms of fat stigma are greater than the purported harms of being fat; in many cases fat stigma stimulates the same inflammatory responses in the body that have been presumed to be from carrying excess fat.[56] Being fat is not inherently unhealthy; being constantly assaulted by fat stigma is. (And to be clear, even were fat and health as closely tied as many purport, it would not be a justification for fat phobia or for an interventionist response that presumes that health and healthism should motivate all decisions.)

And fat stigma is harmful to everyone. You may recall from the chapter on energetic stressors that a severe energy shortage (almost always from intentional caloric restriction) can lead to the total loss of the menstrual period, a condition known as amenorrhea, or, more precisely, functional hypothalamic amenorrhea. While this is partly explained by the fact that the body is conserving maintenance effort while taking away reproductive effort, that's only part of the story. In one study led by the neuroscientist Dr. Kristin Sanders and the gynecologist Dr. Sarah Berga, women with functional hypothalamic amenorrhea were compared to healthy controls in a test that measured their heart rate, glucose, and cortisol responses to exercise.[57] The two groups of women did not differ in their body mass index, psychological profile, or exercise profile. Yet when subjected to an exercise test—a twenty-minute cycling effort at 70 percent of their VO2 max (that's the maximum rate of oxygen consumption during exercise)—there were several differences in women's responses.

Compared to the control group, women with functional hypothalamic amenorrhea experienced a smaller increase in heart rate, greater glucose consumption, and a higher cortisol response. The cortisol difference between these groups suggests that women with functional hypothalamic amenorrhea have either an increased responsiveness to or a greater degree of psychosocial stress. Some papers have shown that women with functional hypothalamic amenorrhea are more likely to have worse coping skills, while others have shown that exercise is often used as a coping strategy—one that only worsens the original amenorrhea.[58] Even in cases in which there is an initial energetic explanation for changes to the menstrual cycle, psychosocial stressors can inhibit a person's ability to return to their typical pattern.

Living in a culture that promotes thinness, and participating in a sport that valorizes it, could itself be a significant stressor to a person of any size. Navigating body-size expectations, oppressions, and discrimination is something that affects all people, but in most cultures, women, gender-diverse people, and gay men tend to face the strongest social pressure around body size.[59] Recent research has found that some people are especially vulnerable to fat-phobic messaging, which takes a great toll on their mental and physical health. For many, the toll comes in the form of clinical or subclinical eating disorders; for others, it leads to HPA axis disruption that, if they are menstruating people, can lead to missed periods and even infertility. People with a history of eating disorders are not only more likely to have fertility problems but are more likely to experience sexual dysfunction (such as anorgasmia and pain with sex).[60]

The relationship between psychosocial stressors and the menstrual cycle is complicated. While everyday hassles may not have a meaningful impact on one's menstrual cycle, stressors that derive from systemic injustices or profound early adversity, like discrimination or sexual abuse, often will. The pathways through which stressors can affect the menstrual cycle—such as the inflammatory and glucocorticoid systems—are varied and interact with one another in complex ways. One thing we do know is that being believed, having a support network, and having appropriate resources to help handle our stressors can lessen their physiological effects. One study that measured the perceived stress and cortisol concentrations of Israeli infertility patients enrolled half of the patients in a cognitive behavioral therapy intervention.[61] Roughly half of the women receiving the therapy intervention (eight of seventeen) became pregnant. In the control group, only one of the eight became pregnant.

While many of the perturbations of the menstrual cycle caused by psychosocial stressors are reversible when those stressors are allayed, we should also pay attention to the harm the stressors can cause by making people feel that their bodies are behaving differently than they expect, perhaps compounding the original psychosocial stress. The more we clarify these roles and relationships and educate people about how stress affects their bodies, the more we can provide comfort to help them weather the storm.

HOPE AND HEALING

Nearly thirty years after Anita Hill testified that Clarence Thomas had sexually harassed her, psychology professor Dr. Christine Blasey Ford testified that Supreme Court justice nominee Brett Kavanaugh had sexually assaulted her in high school. In fact, Dr. Blasey Ford has expertise in the area of trauma and posttraumatic experiences. One of Blasey Ford's most highly cited papers is a collaboration on posttraumatic growth after the September 11, 2001, terrorist attack in New York.[62] Blasey Ford and her colleagues administered an internet survey nine weeks after 9/11 and again about six and a half months later. They found that those who had more trauma symptoms eventually experienced more posttraumatic growth. That is, the symptoms some people experience are part of an adaptive complex to help, in the words of the authors, "cognitively metabolize" the trauma. While posttraumatic stress disorder is a lingering, perhaps maladaptive response, posttraumatic stress symptoms are often important to the healing process.

The paper also reports that early denial of trauma also seemed to be associated with traumatic growth. Early denial can be a protective mechanism, as it gives the traumatized person more control over the rate at which she experiences her hurt. Some of the healthiest and most adaptive means of recovering and even growing after trauma include experiencing significant symptomology—allowing yourself to feel your feelings—yet exerting some control over the rate of those feelings through some avoidance and denial. Perhaps this is why so many survivors of varying types of abuse do not label it as such in the moment.

In Blasey Ford's testimony at the Kavanaugh hearings, she often used her considerable scientific knowledge to protect herself from the intense experience of testifying about her earlier sexual trauma. In recalling her assault, Ford offered the following description:

> FORD: Indelible in the hippocampus is the laughter, the laugh—
> the uproarious laughter between the two, and their having fun
> at my expense.
>
> LEAHY: You've never forgotten that laughter. You've never
> forgotten them laughing at you.
>
> FORD: They were laughing with each other.
>
> LEAHY: And you were the object of the laughter?
>
> FORD: I was, you know, underneath one of them while the two
> laughed, two friend—two friends having a really good time with
> one another.

"Indelible in the hippocampus is the laughter" is a line that I think about a lot. I think about how deliberate and calm it sounded—I can hear her voice in my head when reading the transcript. The hippocampus is a part of the brain that helps consolidate short-term memory into long-term memory. But it is also part of the limbic system and therefore is important to emotion and the threat response. When we experience a significant stressful event, the consolidation of short-term memory into long-term memory is going to be affected. It could compromise the memory, intensify it, or make it go away altogether. A missing memory could reemerge years later, or an intensified memory could lead to constant intrusive thoughts. Inflammatory mechanisms appear to be at least partly implicated in these processes. The laughter at Dr. Blasey Ford's expense likely became an intensified memory, a moment of powerlessness and fear translated into something she could not forget.

"Indelible in the hippocampus is the laughter" is indelible in my own hippocampus. Perhaps it's because that line distills, in a pure and simple way, how so many of us have memories that we would love to forget but cannot (at least not yet). I think it's important to notice that psychosocial stressors can lead to positive adaptations: we can learn coping skills, and we can learn to extinguish our fear with repeated exposure. It means there is hope, even perhaps for those experiences that feel indelible. While early experiences may sometimes have long-lasting effects, most people, when the stressor is removed, can eventually return to their range of typical.

Resilience is a major feature of the human condition. And healing in some form, to some extent, is common. We can look at the psychosocial stressors of our lives, at how they may activate our HPA axis, disrupt our menstrual cycles, or otherwise wreak havoc on our bodies. But we can also notice the adaptability and flexibility of the body to restore and transform, to adjust and respond, particularly when we have access to the kinds of money, social support, and social services that ease the most significant social stressors, such as racism, sexism, poverty, homophobia and transphobia, ableism, and fat stigma. The solution to negative menstrual experiences requires communities of care and structural change, rather than an individualist approach in which each menstruating person tries to handle their symptoms or fertility struggles on their own. I do not want a future where the care I receive is dictated by my ACE score or cortisol reactivity or whether my health provider knows about my early sexual trauma. I want a future where we look upward from these mechanistic explanations for how experiences get into the body and notice the structures that caused them in the first place.

CHAPTER 6

The Future of Periods

WHEN HORMONAL contraceptive pills were first introduced to the public in 1960, these pills were initially packaged in a bottle, like other drugs. A few years later, Ortho-Novum was the first to create the circular dispenser that so many of us are familiar with: twenty-one days on, seven days off.[1] This dispenser gave a sense of temporality to periods, as they occurred in a regular fashion every few weeks. The "off" week was designed by pharmaceutical companies to create a menstrual period because they felt patients, pharmaceutical executives, and religious officials would find hormonal contraception more acceptable this way. Experiencing somewhat regular menstruation is also a major way people know they are not pregnant.[2] Though menstruating people have for decades been hacking their own contraception to avoid periods around certain life events, such as vacations or athletic competitions, it wasn't until the turn of the twenty-first century that pharmaceutical companies began to sell hormonal contraceptive pills that explicitly skipped placebo weeks in order to decrease the frequency of menstruation. By the 1990s, menstruation could also be suppressed by other injectable or long-acting implant forms of hormonal contraception, though suppression was initially seen as a side effect, rather than selling point, of these methods. Pharmaceutical companies assumed for a long time that menstruating people wanted to bleed.

If anything, the introduction of the birth control pill and its off week did more to entrench the idea that periods should occur with a certain regularity and frequency than menstruation biology ever could. There are many points in a person's life during which they are not menstruating, including their premenarcheal years, their postmenopausal years, during pregnancy, and often for many months of lactation. As I've shown throughout this book, the menstrual cycle can vary substantially, leading to less frequent or even absent menstruation, for a number of reasons. The life of a menstruating person is rarely an inevitable march of four hundred to five hundred continual and regular menstrual cycles.

Chemical menstrual suppression, like hormonal contraception, represents the next step of what the historian Dr. Sharra Vostral calls "technologies of passing." Menstrual management products were the first "technology of passing" in that they allow a menstruating person to move through the world as though she is not menstruating.[3] Tampons make it possible to wear bathing suits and go swimming; all forms of menstrual management products decrease the risk of bloodying clothes, furniture, and bedsheets. Menstrual suppression technologies are a logical next step in the quest to gain customers by pharmaceutical executives, but it also seems like a good idea to those looking to survive in hustle and productivity cultures that leave less and less room for experiences like menstruation, not to mention those for whom eliminating menstruation would help affirm their gender. While the acceptance of menstrual suppression technologies was initially quite low, acceptability has increased dramatically over the past several decades, in no small part due to the advertising of pharmaceutical companies and advocacy by vocal physicians.[4] And the increased accessibility to menstrual suppression technologies are part of what we need in our period (or for some, period-free) future.

I'd love to consider a broader period future: one that also creates more room for menstruating bodies. Given the attacks on reproductive justice, mass shootings, forced detransitions, and our ongoing climate crisis, I struggle often to imagine a just future, let alone to let my children out of my sight. Yet we need these dreams if we want to have any hope of getting a better present and a better future. We need to imagine a future where we acknowledge that we are humans with bodies that need attention and love; that the needs of bodies are all different; that our minds are housed in these bodies and are better off when we don't ignore the house. More than self-care or body positivity, I am advocating for the radical (but not new or original) idea that humans deserve dignity and that dignity means not only accommodating but celebrating and noticing all people.[5]

I would argue that menstrual liberation is about becoming more visible as menstruating people while working to ensure that our needs are considered alongside our community's and planet's needs. One important step in pursuing this path is to shift how we frame the questions we ask about menstruation. As the botanist Dr. Mary O'Brien has powerfully argued, when a scientist seeks to offer a risk assessment on a given problem, it involves adopting a particular cultural frame: how much harm a given individual, community, or ecosystem can handle.[6] So when it comes to menstrual concealment and management, risk assessment asks: What concentration of volatile organic compounds in menstrual pads is too much? How long can someone wear a tampon before developing an infection? How many disposable menstrual products can a landfill take before phthalates leach into the groundwater? Or with menstrual suppression: What causes the fewest side effects alongside the highest efficacy for the most people?

O'Brien points out that asking these kinds of questions, rather than alternatives assessment questions, "is to contribute to the currently dominant, but suicidal, assimilative capacity approach and practices of our society."[7] Alternatives assessment questions take a different perspective. Pursuing menstrual liberation, one could ask, what alternatives do we have to the creation of menstrual products that contain harmful endocrine-disrupting chemicals? What alternatives do we have to a life so tightly scheduled or privacy so hard to come by that menstruating people need to wear these products for twelve or more hours at a time? What alternatives do we have to suppression technologies that are not universally effective or tolerable? In *Pollution Is Colonialism*, Dr. Max Liboiron reminds readers that there is no "away" where we can put plastics where we are not ultimately shoving harm on someone else.[8] Plastics aren't going anywhere. Likewise, there is no away where menstruating people can put our messy, leaky bodies—and, frankly, people in power will continue to notice and deride their existence whether or not those bodies conform to their version of professionalism and civility. A risk assessment perspective would ask what menstrual concealment and suppression technologies could support our autonomy with the least amount of harm. An alternatives assessment perspective would ask whether concealment and suppression are the only options on the table.

My own alternatives assessment, my period future, imagines a world where we have ready access to single-stall bathrooms with plenty of menstrual management supplies that are safe and easy to use; where we don't have that feeling of a completely saturated pad and realize "Oh shit, I have to be somewhere in two minutes"; where we know that the products we can readily access and are effective for us do not harm us nor hasten climate change. I even imagine rooms where people could hang out from time to time with a heating pad or some ibuprofen. Bio breaks would be built into the day and care work would be the basic expectation people have for each other at home, at school, at work, and out in the public. What I am imagining is a world where it is as unremarkable for someone to openly carry a tampon as it is to carry a hair clip and where discussing the care of our bodies does not label us weak.

If this sounds like disability justice, that's because it is. In her book *Care Work: Dreaming Disability Justice*, Leah Lakshmi Piepzna-Samarasinha writes about the "crip skills" those with disabilities have developed to make their way through the world and how those skills are what make it possible to problem-solve, support each other, and change environments.[9] These are skills all of us with bodies need, for two reasons. First, disability is a liminal category people can move in and out of over the course of their lives; we can break an ankle, develop chronic illness, need recovery time after surgery. Second, our present and future get better when we consider accessibility as a foundation for justice. As Piepzna-Samarasinha declares, "When I think about access, I think about love."

If my environment were more accessible, more accommodating of menstruating bodies, would I still want to excise periods from my life? I don't know. I know that I am exhausted by the ways I have to manage my body to fit in this world. As I've grown older, I've found myself less interested in changing myself to fit the world and more interested in making the world accept and make room for me and my friends, colleagues, and loved ones.

In this final chapter, I'm going to discuss lingering issues around menstruation, largely around the inadequacy of the medical research apparatus to devote resources or thinking to the needs of menstruating people. Menstruating people need safe ways to manage periods, including stopping them; they need treatment and care for their chronic illnesses; they need a complete reenvisioning of how they fit into the world, especially in public. Perhaps period futures can be messy: incommensurable at times, changing, flexible. Perhaps we need to consider the possibility that there will never be a menstrual alternatives assessment or technology that is perfect but that there is something to be gained by trying.

MENSTRUAL SUPPRESSIONS AND MANIPULATIONS

Most menstrual suppression technologies are varying types of hormonal contraceptives, which are not nearly as well tolerated by menstruating bodies as most of us believe. Across multiple studies, about half of people on hormonal contraceptive methods discontinue them. Even those who do stick with hormonal contraception often experience unwelcome side effects, which they endure as an acceptable cost in order to avoid getting pregnant or menstruating. Many groups are invested in menstruating people staying on hormonal contraceptives, including pharmaceutical companies, those who fear teen pregnancy, and those interested in global population control. But it's possible menstruating people are not always as invested themselves, at least in the management and suppression technologies as they currently exist. According to a recent Cochrane review—effectively, the gold standard in health care if you are trying to assess quality of evidence—direct, in-person counseling, the most common intervention for improving the continuation of hormonal contraceptives, does not increase the rate at which people choose to stay on hormonal contraception.[10] In the papers they sampled, anywhere from a quarter to one-half of those on a given hormonal contraceptive regimen discontinued their use over the study period. One recent study comparing self-reported continuation rates to actual pharmacy claims suggests people may overestimate how continuously they use hormonal contraception.[11] People skip a month here or there because they forget to get their prescription in time, because the prescription is expensive, because they aren't having potentially conceptive sex, or because they don't love how the hormonal contraception makes them feel and need a break from it.

Hormonal contraception, especially shorter-acting forms like pills, rings, patches, and injections, are a hassle, and users often report side effects, such as loss of libido, weight gain, vomiting, dizziness, and depression, as well as amenorrhea, irregular bleeding, and heavy bleeding. Two studies have reported some improvement in continuation among users with adverse side effects who received counseling, but the certainty of the finding was weak.[12] Note that the goal of these studies was to figure out how to keep people on hormonal contraception who were suffering serious effects. The fear of pregnancy—particularly the fear of the wrong person getting pregnant (for example, a teenager or a brown or Black person)—motivates the continued use of hormonal contraceptives that cause harm to about half of the people who try them.[13]

Significant side effects and high rates of discontinuation also plague the levonorgestrel-containing intrauterine device, or hormonal IUD. One study that examined the experiences of 161 women who had the hormonal IUD inserted at one hospital in the United Kingdom found that almost half of them had their IUD removed due to side effects, including "bloating, headaches, weight gain, depression, breast tenderness, excessive hair growth, greasy skin, acne and sexual disinterest."[14] This finding is particularly striking since these women were great candidates for the hormonal IUD: they had had a gynecological exam before having it inserted and, in most cases, also had hysteroscopic assessment of their uterine cavity to make sure they didn't have fibroids or other lesions that could complicate their experience.

In a study interviewing physicians who administer hormonal contraception, respondents were less than understanding when patients requested early removal of the IUD. Physicians in this sample were often frustrated when patients were dissatisfied with their IUDs for any reason. Intent on getting as many people as possible to use them, a physician from the study confessed: "I don't try to influence women's decisions, but I do try. Like I don't want me to be the person making the decision, but I do want to guide them to make a good decision for them. But I usually say it's my favorite method. . . . And I usually say that it's our most effective method and it's very easy to put in." When patients asked to have their IUD removed, physicians often discouraged them by requesting that they keep it in for a few more months to see if symptoms change. While many physicians emphasized the importance of patient decision-making, others only grudgingly ceded to patient autonomy. Others expressed disappointment or disagreement with their patients. These coercive stances run counter to the broader goals of reproductive justice.

Gynecologists often see their family-planning role as one in which they stop as many unplanned pregnancies as possible and in particular get as many people on long-acting methods (like the IUD, Depo-Provera, or implant) as possible. Psychologists Dr. Patrick Grzanka and Elena Schuch point out that this is part and parcel of the original goals of fertility control, which was directed first and primarily at white women and pitched as an opportunity for personal agency. However, different people are going to experience and be pressured differently in long-acting contraception usage, leading to a kind of "conditional agency" depending on their race and gender.[15] The overfocus on how successful a product is in preventing birth, over side effects or one's ability to stop and start the product at will, is how physicians are taught to recommend contraceptive options.[16] This limits patients' abilities to generate their own priorities to inform their choices.

Beyond discouraging patients from removing IUDs, and beyond even how hard it can be to find a physician who will remove one, physicians also discount the pain caused by IUD placement.[17] For instance, the science fiction author Monica Byrne recently shared on Twitter the story of a friend's unanesthetized IUD insertion.[18] The replies that followed demonstrated how frequently menstruating people are discouraged from using local or systemic anesthetic during multiple procedures—not just IUD insertion but IUD removal, hysteroscopy, and endometrial biopsy.[19] Research on pain management during these procedures is hard to come by; the most recent meta-analysis I could find on local anesthetic for IUD insertion was from 2018 and included only eleven studies.[20] In the Twitter thread, people shared stories of health insurance companies that had refused to cover better methods of IUD insertion, such as insertions guided by ultrasound rather than by feel.[21] Many of the Twitter posts describing insertions where the respondent was not offered or was denied pain medication described the pain as worse than childbirth. Many of these respondents were also horrified by their doctors' dismissive responses. As Byrne responded to one tweet, "It's like they build gaslighting into the pre-op." Many respondents, including the original sharer, expressed gratitude toward and/or approval of the IUD, even though it had also caused them significant pain and even trauma. Many people need both contraceptive and menstrual suppression technologies, yet these technologies can come with a steep personal cost.

Safe and effective contraception and menstrual suppression are necessary medical technologies for many people, especially as access to abortion continues to erode. Yet the main method of reversible menstrual suppression seems to be tolerable, at best, about half the time for those who try it. As far as I know, there are no methods being tested right now that do not require hormonal manipulation in order to reversibly suppress the period. I cannot imagine such a low level of research and development were these products that primarily affected cisgender men. And we need to be able to critique the systems that have led to inadequate contraceptive solutions for so many, while continuing to improve access to what we have. This is not about discarding hormonal contraception. It's about working toward something that works better for a greater number of people.

So part of my period future involves pushing structural solutions for the development of feminist technologies to improve menstrual suppression for those who want it. We need better legislation and regulatory oversight on the development of new methods to stop menstruation reversibly. We need tax dollars and other incentives for biotech and pharmaceutical companies to work on novel therapies. And we need to improve the gender ratio in these industries so that more menstruating people are involved in the decisions regarding how money is allocated for research and development departments.

WAITING FOR THE CURE

In addition to developing safe, effective technologies for menstrual suppression, my period future involves developing solutions for many chronic conditions associated with menstruation. There are a significant number of people for whom their period is more than just a hassle. Some are in near-constant pain, and not just during their periods, from endo, fibroids, or adenomyosis. Others bleed so much that they see little point to leaving their house during the menstrual phase. People suffering from heavy menstrual bleeding can become so severely anemic that they can barely walk up a flight of stairs without getting winded. Those suffering from premenstrual dysphoric disorder experience debilitating levels of anxiety, depression, and even suicidality in the days leading up to their periods, as well as during the first few days of their menstrual phase.

None of these populations are particularly well served by modern medicine. For instance, as the gender studies scholar Dr. Cara Jones points out, "Those with endo must grapple with both lived experience of endo as a pain disorder in addition to gendered norms that render it invisible in public discourse."[22] We have known about endometriosis for many decades, yet there is little we can do for it. People suffering from endo are frequently told that their main options for the management of their condition are to go on hormonal contraceptives, get pregnant, or have a hysterectomy. Indeed, this was a common refrain I heard from clinicians and researchers (repeated with frustration, not agreement) at a menstruation science conference held in 2018 at the National Institutes of Health. Yet none of these options actually reduce endo or endo pain for most people. Instead, they provide the illusion that one's clinician is doing something when many know these recommendations are not only problematic but ineffective.

Consider the first option, hormonal contraceptives. According to a meta-analysis of studies exploring the relationship between endometriosis and hormonal contraceptives, people on hormonal contraceptives have a mildly reduced risk of experiencing endometriosis. However, as the authors of the study point out, since hormonal contraceptives provide some relief of endo symptoms, the "reduced risk" might be an artifact of delayed diagnosis due to suppressed symptoms.[23] What's more, women with a history of hormonal contraceptive use who do not have children end up with an increased risk for endometriosis later in life (this risk is inverted for women who have children, for whom hormonal contraception appears to be protective).[24]

The second option, having a baby, does seem to have a substantial effect. Several studies have shown both that having children lowers one's risk of endometriosis and that the higher hormone levels associated with pregnancy may lessen existing lesions.[25] Yet endo causes infertility in at least a third of sufferers, so having a baby is not going to be the easiest cure. More importantly, it is not exactly an option for those who do not want to bear children, not to mention an outrageous reason to conceive a baby. And in a recent sample of women being surgically staged for endometriosis (often considered the gold standard for diagnosis), women who had given birth were more likely to have adenomyosis, a condition in which the endometrium pervades the uterus's outer muscle.[26] Morally and scientifically, having a baby is not a viable solution for endo sufferers.

We are left, then, with the final option, a hysterectomy. In *Girl, Woman, Other*, Bernardine Evaristo's 2019 Booker Prize–winning novel, Carole experiences debilitating menstrual cramps each cycle. Her pain limits some of her daily activities and, as a Black British woman working in an investment firm, risks limiting her standing at work:

> the only morning she doesn't run is when she's doubled over with period pains for which she takes extra-strength painkillers in order to haul herself to work or risk being accused of taking a monthly sickie
>
>   busted! yes, you *are* a woman
>
>   she even contemplated having her womb taken out to eliminate periods altogether, which would surely be her greatest possible career move, a tactical hysterectomy for ambitious women with menstruation problems[27]

Perhaps Carole is not entirely serious here. Yet we should not minimize how exhausting it is to be in pain, to hide that pain from the world or risk mockery, and to be reminded not only of our gender but of our gender inequality. I know many people whose menstrual pain and discomfort significantly interferes with their lives. Some of them, seeing no other way to end their pain, have had their uteruses removed. I also know menstruating people who did not see the point of holding on to an organ for which they had no use. Yet they (especially if they were or appeared to be cisgender women) experienced significant pushback from doctors who thought their patients' primary concern should be fertility preservation.

Hysterectomy is the most common surgery performed on women in the United States, after cesarean section. Contemporary analyses of nonmedical variation in rates of hysterectomy reveal some troubling trends: hysterectomies are more often performed on Black women than white women, and male gynecologists are more likely to recommend them than are female gynecologists.[28] While hysterectomies are described as a nonreversible, permanent solution, the procedure does not guarantee the end of endometriosis, which is defined by the lesions that live *outside* the uterus. Removing the uterus does not remove existing scarring, which can still lead to a recurrence of endometriosis, requiring yet more surgery. In fact, more than half the time women get a hysterectomy to treat endo, it comes back.[29]

There are also several major, often undiscussed consequences of hysterectomy. Women who undergo a hysterectomy—nearly half of all women in the United States—face increased risks of heart disease and hip fracture from the loss of ovarian hormones.[30] These risks are greatest for women who also have their ovaries removed, but even women who retain their ovaries tend to experience an earlier menopause, as well as an increased risk of dementia and Alzheimer's disease. They are also at greater risk for pelvic floor dysfunction and urinary incontinence.[31] I recently told a friend about the pelvic floor consequences of hysterectomy, and she said, "Oh yeah, all my friends with early hysterectomies pee themselves now."

One study of outpatient hysterectomies showed that white women are more likely to be offered less invasive methods for hysterectomy than Black women, which allows them to recover faster and often experience fewer complications.[32] Epidemiologist Dr. Whitney Robinson, who led this study, told me that it was originally assumed that this disparity had to do with the assumptions around Black women being more likely to have complex cases because of their weight or having more progressive disease. If that's the case, then as the field moves toward less invasive methods, Black and white women's noninvasive surgery rates should still increase over time, even if the proportion of Black women with less invasive methods is lower. Instead, there is a significant lag in adoption of less invasive methods being used for Black women. As Robinson says, "Just when you are thinking about the options you have, you realize peoples' options are not the same."

For menstruating people who are men, are masculine of center, or are otherwise gender diverse, suppressing the menstrual period can be a crucial component of gender-affirming care. While testosterone therapy reduces the frequency or even eliminates menstruation in most people who use it, it can take many months, sometimes up to a year, until users stop menstruating.[33] Many transgender men and masculine of center folk also choose hysterectomy—indeed, for many years it has been strongly encouraged in the medical community—and few are made aware of the risks described above.[34] The rates of postoperative complications for transgender men getting hysterectomies appear to be comparable, according to an analysis of all hysterectomies recorded in the American College of Surgeons' National Surgical Quality Improvement Database from 2013 to 2016, though postoperative complications are defined more narrowly (infections, deep vein thrombosis) than the kinds of complications that can emerge over a longer span of time.[35]

As more masculine-of-center people are choosing pregnancy, lactation, and chestfeeding, we need to ask ourselves if the push to give hysterectomies to this population is about reducing gender dysphoria or about policing gender/sex. Public health professor Dr. Michael Toze points out that medical encouragement of hysterectomy toward transmasculine people stems from two interrelated cultural issues: our longstanding desire to fix the inherently pathological female body and our collective discomfort with the idea of a pregnant man. When people choose to view female and transmasculine bodies as much the same thing, they end up being painted with the same broad brush and subjected to the same sexist discriminations. As of this writing, several countries require sterilization in order to legally change gender from that assigned at birth. Others strongly recommend it. Such requirements perpetrate eugenic fertility control on transgender and other gender-diverse people in the same manner that is perpetuated against those who are Black, poor, or disabled.[36]

MAKING ROOM FOR MENSTRUATION

The science-forward understanding of menstrual periods that many Western cultures have developed and popularized has had some beneficial consequences. Menstruating people in these cultures, having received more neutral information about what periods are, are less likely to be surprised or frightened by the experience than in the past.[37] However, it is also plausible that this development explains why those living in Western countries generally report worse menstrual symptoms than many other populations. As we have destigmatized menstruation by focusing on the science, we have also developed a tendency to describe menstruation almost entirely with negative symptomology. There are, for instance, many studies of negative symptoms through the menstrual cycle, but to my knowledge, only one scale has ever been developed to assess positive experiences.[38] If all we ever learn is that menstrual cycles make us hormonal, irritable, bloated, angry, depressed, anxious, or in pain, is it any wonder that's the primary way many of us perceive our experiences?

I also wonder: Is part of the problem the way Western culture hierarchically places mind above body? We all must tend our bodies. Those of us with menstruation, gestation, lactation, chronic illness, and/or disability often have to tend them even more, rendering this hierarchy and separation nearly impossible. A society that ignores the needs of the body will be considerably easier to navigate for able-bodied, cisgender men, who do not have to deal with these additional care needs, and they are less likely to be tasked with the care of other bodies, like children and the elderly. The rest of us are going to struggle with this from time to time. By pretending we don't have bodies, we develop a culture in which workers brag about forgetting to use the bathroom or eat. We also build workplaces with no slack time between meetings, no medical or maternity leave, and two fifteen-minute breaks per day for those who lactate. Case in point: after the birth of my second child, I saw a pelvic floor therapist. She tried to pump during her two fifteen-minute breaks each day, but it just wasn't enough, and by the time my six insurance-supported sessions with her were done, she, a health provider working within the health-care system, had had to stop breastfeeding.

As the anthropologist Dr. Emily Martin points out in her landmark book, *The Woman in the Body*, the way we conceptualize menstruation places it in the "private realm of home and family." This is a large part of what makes being a menstruating person such a hassle when we are in the realms of school and work and why many menstruating, lactating, and/or disabled people are often less excited about a full return to the in-person workplace.

Though Martin's book was published in 1986, her interviews with American women at many stages of life do not read as dated. Martin notes that menstruating people separate out reproduction from the self by describing menstruation as a process, menopause as a stage, birth as a series of signals sent from the body to the self, and labor as something to get through—terms we still use today. Toward the end of her book, Martin shows how this separation of body and mind limits even those activists trying to return power to pregnant people. Over the past several decades, birth activists have sought to bring control to the birthing process back to people giving birth and to demedicalize something that is not inherently pathological. The intentions behind these efforts are often good (to declare my own bias: I gave birth to both of my children at birth centers rather than hospitals). However, these efforts often involve a juxtaposition between what is medical and "cerebral" and what is "natural." As Martin points out, these efforts "occur at the cost of reasserting a view of women as animal-like, part of nature, not of culture."

The idea that menstruation should remain within "the private realm," as described by Martin, is another idea that has impeded even well-meaning activist efforts. As gender studies scholars Dr. Chris Bobel and

Dr. Breanne Fahs have shown, menstrual activism often suffers from a desire to conceal and manage menstruation rather than make room for it in the world.[39] The primary goals of menstrual activism are frequently *product centered*, meaning many organizations aim to get menstrual products into the hands of people who need them. Again, this is a laudable goal! But it continues to reinforce existing hierarchies and the politics of respectability that suggest menstruating people need to conform to existing norms of propriety in order to be deserving of dignity.

People of all genders can feel the need to fit into cisgender male spaces, especially those, like work and school, that were originally built without their needs in mind. This may lead them to hide processes like menstruation or lactation so as not to remind anyone they have a body that requires tending. Compounding the need to fit into spaces designed for and by men is the shame that often accompanies these processes. With menstruation, I often feel like part of the shame is the complete uncontrollability of it. Unlike voiding your bladder or bowels, unlike feeding your body, menstruation is something that happens that you cannot stop and start. It gushes, it wells. Chunks are sometimes slipping out of our bodies, creating significant sensation. All the while, we are expected to behave as though none of these things are happening to us in real time.

I think on the number of times I have been in a meeting that is going late, I know I really need to change my pad, and I can feel my menstrual effluent heavily leaving my body. Yet I attempt a relaxed posture, I shift in my seat, and I hang on as long as I can with a smile plastered to my face. Or when I've been teaching and feel a deep, aching, cramping sensation pulling at my lower body. Yet I keep on talking my way through my slides at the front of the classroom. Part of my brain is occupied with whether, somehow, my pad was poorly placed, and I'm bleeding through it, but I have no covert way to check. Despite the prevailing cultural association between femininity and being emotional and out of control, femme people are constantly holding ourselves in check. Ultimate femininity, as well as the ability to fit into spaces where everyone cares about your mind and no one about your body, is about passing as not menstruating.

Living in a world dominated by this dualistic, male-centered perspective has disconnected me at key points in my life from the world around me. I wonder sometimes if some negative symptoms associated with processes of the menstrual cycle and/or reproduction are in part a response to how out of place we can feel. Perhaps, as the professor of social policy Dr. Val Gillies and colleagues have written concerning bodily processes like sweating and pain, our dualistic understanding casts the body's processes as intrusive to the thinking of our minds: inappropriate and out of control.[40] If our bodies' normal processes are intruding into our minds' work and we've been taught to prize certain workings of the mind above all else, it follows that we might not view these intrusions all that positively. We might try to hide these intrusions from the world.

Then there's the pain. We have been "invested in de-emphasizing the hazards and pains of menstruation as a way to fight the pathologization of menstruating bodies, diminishing the reality of the painful, cyclical, and chronic aspects of bleeding."[41] In doing so, we risk replicating medical models of menstruation. Medicine has a long history of neglecting menstrual pain, of calling people who suffer from it hypochondriacs and hysterics, of considering it both inevitable and beneath medical attention. Even when pain is part of the suffering that goes with a reproductive condition like endo, what largely defines the treatment is restoration of fertility rather than pain management. The only pain management that feels readily accessible when it comes to reproduction is during birth, and much of that pain management, to be frank, also helps to control the birthing person because it renders them immobile or disallows eating, movement, or other behaviors.

Some of the most vulnerable and painful moments of my life have been moments I have hid from the world. After each of my failed embryo transfers, I bled more than I thought possible. I went to work. I taught class. I cried in my office . . . howled, really. I would replay over and over the moment our then seven-year-old found out I was not pregnant after that first failed transfer. They lay on my bed and, sobbing, said, "I don't think I can ever be happy again!" My husband and I each curled around them, and we cried. I bled, and I remembered that moment, and I went to work. Some of these things I did because going to work helped me get out of my grief, or my pain, or my fear. But even as I was using work as a way to not ruminate on what I was experiencing, I remember the distance I felt from myself. What the actual fuck was I doing going to work and pretending like I was not immensely suffering? And how many of my friends and colleagues were doing the exact same thing?

As we think about how to transform our world, we need to ask ourselves: Who made the world we live in? Who benefits from it being the way it is? And what are our alternatives? The limitations of my own creativity led me to look next to speculative fiction. Science fiction and fantasy authors are some of the people most engaged with the creation of new worlds. What happens to periods in their stories?

BODYMINDS AND SPECULATIVE FICTION

The first fantasy novel I can remember reading as a child was *The Blue Sword* by Robin McKinley.[42] It was also one of the first books I read that did not seem intended for a young audience, providing a gateway to adult science fiction, fantasy, and horror. Like many of my generation, I read Stephen King far too young (I still remember trying out phrases I learned from *The Stand* on my father and how his eyes bulged[43]), but I also discovered the works of Arthur C. Clarke, Robert Heinlein, and Anne McCaffrey. What kept me reading, even as I noticed these stories' limits, was the potential to dream of new worlds. There was the absolute refusal to accept that life as it is, is as it should be.

Dr. Lisa Yaszek, a professor of science fiction studies at Georgia Tech, says that speculative fiction matters because it is "the premier story form of modernity, especially technoscientific modernity." Yaszek's own scholarship often focuses on recovering lost voices of science fiction and analyzing the futures that these ignored or forgotten authors put forward.[44] Speculative fiction authors and readers create a space to discuss our hopes and fears about science and technology and their social impacts. Consider the enduring effects of *Frankenstein*, written by Mary Shelley, daughter of the feminist philosopher and author Mary Wollstonecraft.[45] "Shelley sets up a certain template for thinking about how masculine technoscientific intervention could have bad impacts on women as reproductive subjects," Yaszek points out. "That's exactly what we're still grappling with today, right?"

The amount of positional power an author holds in their culture, directly or in proximity, can affect their ability to envision different futures. There are works that are read as period metaphors, like Stephen King's *Carrie*.[46] (Yaszek deadpans as King: "I'm a white guy. And when I hear about periods, all I think of is violence and bloodshed and anxiety and terror.") There are a lot of works from the 1960s through the 1980s written by white women that imagine an away for periods, full suppression, and even womb-like incubators to free women from the reproductive burden altogether.

(One of my favorite authors from this time, Connie Willis, has two short stories that at least partially disrupt the standard away account: "Daisy in the Sun" offers a disturbing story of a girl getting her first period the same moment the sun goes supernova and how, in Yaszek's words, this leads to a sort of "quantum entanglement."[47] And "Even the Queen" addresses generational differences in how people with uteruses consider and manage their bodies, though, like Yaszek, I don't love the ending.[48])

To be clear, this is a period future plenty of people can get behind because of the autonomy it offers over one's reproductive experience! My question is simply: Is the path toward a just future one where the solution is always hoping we can find an away for whatever does not suit our environment? And are we able to follow this path and recognize the potential for dystopia?

Let's start with the away assumed in much science fiction: that we can pass off our reproductive burden and have something gestate for us. Parental-fetal feedback is crucial during gestation, meaning part of our gestational process requires an actual human being whose breathing and blood flow and responses to their environment teach the fetus what to expect when born. I do not think mechanical incubators for the whole of pregnancy are in our future. We are instead likely starting to hit a hard limit on premature survival: recent data show that babies born before twenty-eight weeks have a 53–72 percent survival rate depending on the country, with most who survive having a major morbidity like chronic lung disease, vision problems, or bleeding in the brain.[49]

But let's follow this dystopian idea down the road a bit and say that away is the goal. What happens if a womb that is away can only realistically be created by using another person's womb, through surrogacy? Today, commercial surrogacy is growing in popularity, and the primary consumers are white, well-off women and gay men. This leads to a market that prizes some gametes and some bodies over others, and given how pregnancy can be a life-threatening endeavor, people with significant financial precarity may be more likely to choose this line of work. Philosopher Dr. Amrita Banerjee argues that commercial surrogacy global markets create a "transnational reproductive caste system" that reproduces existing social hierarchies.[50] As an example, the geographer Dr. Carolin Schurr has shown how in the Mexican assisted reproductive technology market, the majority of consumers are white gay American men. In this market, where egg donation and surrogacy are kept separate (meaning, the egg donor and surrogate are different people, such that the surrogate is not genetically related to the fetus), white gametes are privileged over gametes of color, and the lives of the women of color who serve as surrogates are made invisible.[51]

A dystopian future borne out of this understanding of global and racial inequality might imagine a tiered system with options to away one's reproductive burden depending on one's means and status: human gestations for well-off families and mechanical gestations for poorer ones. James S. A. Corey's *The Expanse* series, the first book of which was written in 2011, engages with these hierarchies.[52] Later books have a significant plotline about the difficulties of pregnancy in space and how class (and where in space you were born, which is kind of a stand-in for systemic racism) affects access to a safe pregnancy. Technology can be wonderful, but to imagine a future technological away without trade-offs (remember, risk assessment) for menstrual suppression or for reproduction more generally may never be feasible.

But an away is not our only option for how we handle our uteruses in the future. According to Yaszek, while mentions of periods in speculative fiction largely disappeared around third-wave feminism, they came back "with a vengeance" with fourth-wave and transnational feminism. As gender/sex is being reconceived and reimagined in the dominant public's consciousness, more authors are reconsidering periods. And as authors and readers are clamoring for books with more racial, ethnic, nationality, sexual, and gender diversity, more publishers are finally taking note. Yaszek notes the current emphasis is less on eliminating periods. Just a few examples: Tade Thompson's horror novellas *The Murders of Molly Southbourne* and *The Survival of Molly Southbourne* focus on a girl whose blood (including her menstrual blood) creates a doppelganger that tries to kill her; this phenomenon was brought about because of an experimental medical treatment and is kind of a modern mash-up of *Frankenstein* and *Carrie*.[53] Aminder Dhaliwal's graphic novel *Woman World* imagines a postapocalyptic, and quite funny, future without men (it is also clear from nearly the first frame that women are conceptualized inclusively and are not just cisgender).[54] Dhaliwal engages directly with menarche and menstruation. Sylvia Spruck Wrigley's short story "Space Travel Loses Its Allure When You've Lost Your Moon Cup" has a main character bargain away her chocolate stash and promise not to sing Gloria Gaynor for the sake of managing her period.[55] And A. J. Fitzwater's short story "Logistics" tells the story of a trans man who was in the middle of top surgery when the apocalypse happened. As he manages recovering from a half-finished surgery, he wanders Western Europe looking for tampons, eventually finding community and cause.[56]

As the gender and women's studies professor Dr. Sami Schalk explains in her book *Bodyminds Reimagined*, white able-bodied writers often conceive of utopian futures where problems of race, gender, and/or disability are solved by eradicating difference. For instance, all races combine, erasing race as the means to undo racism, or medical cures abound, erasing disability as a means to undo ableism.[57] Black feminist, transnational feminist, LGBTQ+, and more speculative fictions have a lot to offer our potential period futures even when not touching specifically on them because they show us an alternative to solving problems through erasure. These works deal with bodily autonomy, based on frankly a wider grasp of American and global history and a greater number of potential futures.

In these new futures come new possibilities for concepts and technologies not found in dominant science. Schalk reveals the direct contradiction to dominant, and in this case also Western, science frameworks in the concept of the bodymind, which "insists on the inextricability of mind and body" in opposition to Cartesian mind-body dualism.[58] Reproducing, leaky, wet bodies have different experiences trying to conform to dualism. The bodymind therefore offers a different possibility for what bodies could mean to us now and in the future.

Octavia Butler, whose work Schalk discusses at length in her book, is one who offers a future that grapples with systemic oppression, particularly racism and sexism. While I've read all of Butler's books, I return to *Parable of the Sower*, the near-future dystopian story of Lauren Olamina and Earthseed, the most.[59] Her verses, meant to inspire and teach a new way to survive, are sprinkled through the text. Olamina says famously, "God is Change," which I take to mean that we need to accept change but also shape change. She teaches that everyone is connected, that my survival is predicated on yours, and that we shape our environment as much as it shapes us.

I hear responsibility in these words, and accountability. I hear community. A future I want to work for is one that looks something like this.

EPILOGUE

WESTERN CULTURE'S toxic trait (nowhere more obvious than in the United States) is pretending big structural problems are in fact personal ones. Within this framework, the locus of control for all things that happen to menstruating people is within them, such as how we manage exposure to synthetic hormones and endocrine-disrupting chemicals, meet biological needs in public, and more.

Doctors, advocacy groups, Instagram health enthusiasts, and mothers-in-law might all have opinions about how menstruating people should be treating our bodies so that they behave in some particular, narrow definition of normal. Most of this advice is a demand for personal transformation in terms of what we eat, how much we move around, and whether we "allow" certain stressors in our lives. Somehow, even though we anthropologists call these factors *environment*, in popular parlance they are all considered to be within an individual's sway.

If you have money, status, or power, then maybe to some extent some of these stressors are under your control. You can pay for someone else to clean your house and expose themselves to cleaning chemicals and dust filled with phthalates; you can pay other people to paint your home and breathe in volatile organic compounds; you can buy a home near green space that reduces your exposure to combustion and noise pollution in a space that could support several families; you can buy factory-farmed produce shipped from thousands of miles away. The choices we make to protect and take care of ourselves operate within a broader system and set of interconnected communities to whom we should feel more accountable. But if you want to, you can pay for a partial "away."[1]

The myth of personal responsibility overstates the ability of any one person to protect herself through her individual choices. I can buy high-quality silicone menstrual cups and chlorine-free menstrual pads. But none of the multistalled bathrooms in my building at work make it easy to change menstrual products or dump out cups. None even have trash cans in the stalls, so we have to carry out our used pads and tampons with bloody fingers. My personal choices about menstrual management products do not change what major menstrual product corporations choose to make, what runoff their factories produce, or how the majority of consumers choose to discard their used products. Ethinyl estradiol, one of the main components of most hormonal contraceptives, is found in fish near wastewater-treatment plants and in groundwater; this synthetic estrogen is more persistent in the environment than natural estrogens made by the body.[2] We know it is harmful to a lot of marine life and to humans, increasing breast cancer and possibly prostate cancer risk.

And while exposure to endocrine-disrupting chemicals is the most obvious way to show the lie of individual choice, we can also look to the effects from energetics, immune challenges, and psychosocial stressors. Cultural pressure to be thin and feminine is not going away, and we have done little in American culture to reckon with the racist and sexist origins of fat phobia; therefore, eating disorders and subclinical disordered eating practices that affect menstrual cycles are not going anywhere just yet. We are still in a literal pandemic, dealing with a disease that not only kills but leads to chronic pain and disability, worsened by individualistic messaging and disinformation. Most politicians seem completely unwilling to enact gun control legislation to protect us from mass shootings. And while there is increasing social pressure for people to not state their most egregious views on gender, race, class, sexuality, or disability at work, compliance and punishment-focused policies to reduce these behaviors have only meant that sexism, transphobia, racism, sexism, homophobia, and ableism have gone underground.[3] My insomnia has worsened over the last five years, and it's not just the beginnings of perimenopause.

When I interviewed Dr. Whitney Robinson for this book, the Tops grocery shooting in Buffalo, New York, where a white supremacist murdered ten people, had just happened.[4] Days later, twenty-one children and teachers at Robb Elementary in Uvalde, Texas, would be murdered. I asked her how she carries on with her scholarship, which addresses structural racism and health, when her heart is heavy. Robinson said our charge as humans on this earth, and as researchers, is to address material needs. She said, "Part of the real work of caring about life is not keeping somebody breathing and on life support, but having them be able to live life and flourish, to care about the bleeding as much as you care about some other aspect of their life."

Bodies matter. How they are treated in the world, how they move through the world, what they can access under systemic racism, sexism, and ableism—all of this matters. I think a lot about how much I experience with my body, or rather my bodymind: how I give my children lingering hugs and smell their hair after every mass shooting, how I clear my feelings with exercise after a tough day, how I enjoy the way the warm spring wind rustles my clothes when I walk the dog. Caring about bodies and how they fit into the world—their material needs—matters.

I refuse to be demoralized. If you have gotten this far in this book, it means I have a buddy, a coconspirator. We can be accountable to each other, we can imagine new futures, and we can work toward liberation. What I want for you more than anything is to abandon the idea that you are wholly responsible for your menstrual health and that changes to your menstrual health are your problem and yours alone. Our culture, our environment, the ways in which our various identities are targeted for harm, and our lived experiences shape menstruation. If we are to resolve issues that occur inward, we must direct our efforts outward. A liberatory menstrual future is one where we have the strength together to address power structures and ingrained cultural beliefs. It's great to abolish the tampon tax and to ensure that everyone has access to menstrual products. But remember that many of the interventionist practices I've described around menstrual health in this book attempted to address the outer veneer of a problem in a way that left the real problem intact. Interventions can leave in place the underlying systems of colonialism, extraction, and poverty that led to the problems in the first place.

A liberatory period future does not in fact focus just on periods and period experiences. It casts its eyes upon the power structures that stigmatize and marginalize menstruating people.[5] When we are able to reveal that those same power structures shape our experiences of menstruation through the mechanism of psychosocial stressors, it gives us a target and the beginnings of a solution.

I hope we can start with three steps toward a liberatory future. The first step involves acknowledging and seeking to repair the harm caused by the menstrual stigma that pervades dominant science. The second step requires that we identify, and then involve, the parties and communities affected by menstruation: this includes communities impacted by the creation of various menstrual suppression and management technologies, from those who live near factories, to the janitors who clean bathrooms containing menstrual waste, to the people who need these technologies to get through the day. The third step is that we have to think through and collectively change the structures that continue to stigmatize periods through activism in our local, state, and federal governments; in our workplaces; and in our schools and public spaces.

The first step, acknowledging and seeking to repair harm, entails learning the history of menstrual management, menstrual suppression, and broader gynecological science. Contraceptive and menstrual suppression technologies exist today because scientists created "fertility studies" in places where contraception was barred, concocted doses of oral contraceptives that they knew were dangerously high, and tested formulations on women of color.[6] These drugs and technologies, despite their harms, enabled greater decision-making and control over reproduction and menstruation for countless people. These issues are not magically reconcilable through a history lesson and an apology.

We need to consider what it would look like to acknowledge what happened, listen to the affected parties and communities, and determine what it might take to repair some of these harms. We have modern examples: there have been commendable efforts to educate the world about Henrietta Lacks, the Black cervical cancer patient whose cells were collected and used for science, without her consent, and later became the first and most important immortalized cell line in history. These efforts involved listening to the Lacks family, recounting the story of Henrietta's life so that we may honor her memory, and providing financial restitution.[7] In my experience, many people have a defensive reaction to the idea that we should engage with these histories. "So what, now we can't use contraception? Or now I have to feel bad about the pill?" But that's not it. We don't have to cancel the pill, disposable pads, or the hormonal IUD. We do need to consider what alternatives assessments, and compromises, might look like that include the needs of more people.

The second step on the path to true intersectional feminist technologies for menstrual management and suppression involves moving forward differently than we have in the past. We need to involve affected parties and communities from the beginning when conceiving of and designing drugs, products, and technologies. Feminist technologies are defined many ways: technologies that are used by women (and, I would add, all historically marginalized genders), that improve the lives of women, that empower women, that allow women to "lead their lives" in a patriarchal society, that align with feminist goals, or that help change technosocial systems of patriarchy.[8] The creation of feminist technology cannot, however, be focused only on the end user; rather, it must adopt a broader transformative justice approach. In learning from Liboiron and O'Brien: if a menstrual management product works great for a subset of middle- to upper-class consumers but the factory that makes the products spews toxic chemicals into a rural community or leads to overflowing landfills of nonbiodegradable plastics, we cannot say that it "works." If a menstrual suppression technology is tolerated by half the population that tries it and some portion of the other half is cajoled, counseled, or guilted into staying on it, we cannot say that it "works."

From the beginning, we need to ask: How will our environment be affected by these technologies?[9] Will we be putting synthetic hormones into rivers? Will we be creating by-products that go into the ground, into landfills, into vulnerable communities? Will these products and technologies accommodate the many living styles and bodies of people who menstruate? How will medical staff be brought into the process to learn how to partner with their patients to decide on the right course of action? And if these parties have needs that lead to incommensurable difference, how will we decide together which path is the right one to pursue?

I left the hardest step for last. We cannot continue to create management and suppression technologies that merely aim to help menstruating people pass as nonmenstruating people, though it can be one of the goals. We need to think about the broader power structures that make people feel they need to hide menstruation; that increase negative symptoms; that create financial, social, and personal distress. We need to think about the structure of many peoples' work and school schedules and how they can be reformed to accommodate living, breathing, excreting, eating, gestating, menstruating, lactating, sensing, feeling, hurting bodies. What the world does to our periods is so much bigger than what our periods do to the world. We are not polluting. We are not mindless zombies driven by fluctuations in our hormones. To the extent that we frighten nonmenstruating people, it is only because we remind them that we all move around the world in these fluid-filled sacks that require care and attention.

Menstruation is ubiquitous and intersects with other social injustices. Migrants, refugees, and prisoners menstruate. So do schoolchildren and factory workers. Those of us who menstruate will do it upward of four hundred times in our lives, for five to seven days at a time. This works out to about 20 percent of our postpubertal, premenopausal life. I have menstruated while conducting fieldwork, and I have menstruated while on strike.

As this is a book that covers the science of periods, I want to bring in science here as I close. We have lost time, and opportunity, and creativity to the way we think about periods. But I also see a fascinating future in the people who study periods today. More researchers are moving beyond paradigms of cleanliness or uselessness and toward paradigms of healing and centrality in how we conceptualize periods. With more and better research on periods, we might be able to better grasp the range of variation across all sorts of human environments and leverage that to understand whether and when menstruation is important to a body. With more funding and a more diverse set of researchers, we can ask completely different questions.

In *We Do This 'til We Free Us: Abolitionist Organizing and Transforming Justice*, abolitionist Mariame Kaba paraphrases the work of liberal evangelical Jim Wallace, explaining, "Hope is really believing in spite of the evidence and watching the evidence change."[10] One of Kaba's most famous phrases is "Hope Is a Discipline," meaning to do transformational work is to choose hope every day, to be open to new knowledge, to believe we can work toward a better world. Over the course of the last few decades, our concept of reproducing bodies as fragile has changed. Our understanding of the agency of the uterus has changed. Change is the operating principle, but again, as Octavia Butler shows us, you have to shape it.

I hope you read this book everywhere—in libraries, on public transportation, in front of loved ones and strangers. I hope the next time you pick up a tampon or pad, you feel you can carry it to the bathroom rather than stuff it up your sleeve. I hope the next time you need to purchase menstrual management supplies, you buy a little extra for someone who needs them and consider the effects on the land around you. And I hope the next time you talk to your doctor, you make clear your priorities and needs. I hope you will not let anyone, especially legislators and bosses, tell you that something you do for a quarter of your adult life isn't important. I hope you become part of creating a just period future that practices community, responsibility, and dismantling oppressive structures.

ACKNOWLEDGMENTS

WHEN I was a kid, I couldn't decide if I wanted to be a doctor or a writer, so I became an anthropologist, which is a little bit of both. For all the ways I have been frustrated by the systems that corporatize and casualize academia, I have also continued to be utterly in love with a life where I get to have brilliant colleagues, teach brilliant students, and pursue research that I find important. I cannot express deeply enough how many people have pushed me, inspired me, made me want to be better, over the course of writing this book. If you find things in this book that you love, know that you love them because I know really cool people.

This book should have been done in two years and instead was done in four. Yet looking back on how much I grew as a scholar, I am grateful the writing and rewriting took as long as it did. I needed that time. So first I want to thank Seth Mnookin and Tom Levenson at the MIT Science Writing Program for getting me affiliate status during my sabbatical, without which I would not have had access to the libraries or a lactation room.

I met Alison Kalett, my editor at Princeton University Press, back in 2012 or so. She started trying to talk me into writing a book back then. (I hope she doesn't regret her decision.) Alison, thank you so much for your cool, collected approach to editing, for championing this work, for tolerating my constant and terrible jokes. In ways big and small, you have been in my corner. This book would not exist without you.

Thank you also to my developmental editors Amanda Moon, James Brandt, and Thomas LeBien of Moon & Company. How am I supposed to go back to academic writing without your kind and constructive support? Thank you for helping me with pacing and forcing me to kill long-winded passages about helminths. Hallie Schaeffer at Princeton University Press: you were so conscientious and helpful at many crucial moments of this process! Many thanks also to those who worked on copyedits, page proofs, and of course the publishing and marketing of my book. Because of when acknowledgments get written I don't know you all yet. Yet you are integral to this process and I am so grateful for your work.

There were so many times when I was writing this book that a librarian saved my butt, and most of the time, I never got to know their names. Abigail Goben at the University of Illinois Chicago is one of the few I can name who helped more times than I can count. I am also fortunate that the University of Illinois Urbana-Champaign library is one of the best in the country. I am especially grateful that they were able to track down a conference proceeding from the 1980s that made the opening pages of chapter 3 possible.

I interviewed a number of people for this book, not all of whom ended up in its pages. Thank you so much to Roger Pierson, Angela Baerwald, Jerrilyn Prior, Jodi Flaws, Tory Eisenlohr-Moul, Alanna Nissen, Lisa Yaszek, Ruby Mendenhall, Lilia Cortina, Jamie Jones, Sarah Berga, and Whitney Robinson for your insights and your scholarship.

So many people chatted with me on Twitter, shared their thinking, or shared articles of interest with me: Ricardo Loret de Mola answered random questions; Yasmeen Hussain supplied sexual selection papers I had not seen; Ambika Kamath was generous in conversations on gender-inclusive language in biology; Jessica Greenberg, Jenny Davis, Jessica Brinkworth, Karen Rosenberg, Petra Jelinek, Sarah Richardson, Katie Lee, and Emma Verstraete sent ideas and links and articles. Bryana Rivera provided crucial research assistance for the first draft. Thank you also to Lara Owen, Luisa Rivera, and Alaina Schneider for providing expert reviews on a few chapters where I could tell I needed more direct feedback: they delivered. Thank you to my multiple anonymous peer reviewers (and to the ones who didn't like my coarse language: I kept most of it).

So many friends helped as well: thank you to Ed Yong, who gave me great advice when I needed it, and thank you to Liz Neeley for making a great playlist that I queued up on more than one occasion when I needed to be writing from spite. Thank you to Felisa Reynolds for Tuesday yoga, then Tuesday Zoom yoga, then Tuesday talk for an hour and then fit in a little yoga. And thank you, Jessica Birkenholtz and Lara Orr, for the cheerleading and the article links.

To the entire Clancy Lab, past and present: thank you for listening to me talk about this project for years, for making me smarter and better, and for teaching me so much. Thank you for embodying the values of feminist science, for working to change the world, and for inspiring me every day.

I know we aren't supposed to talk about these things, but having a book advance meant that when our car was totaled, we were financially okay. When Princeton University Press provided me with an equity grant to fund several trips away from my family so that I could write without cluster feeding or overseeing pandemic Zoom school, it was the difference between this book getting done and . . . not. (Shout out to Kacey and Stephanie at the Bee House, the Airbnb where many chapters were written!) My husband makes more than me—he's an engineer and they are valued way more than anthropologists—and that has fueled my courage more than once over the years when I have taken risks that I thought might cost me my job. These are real, material benefits that authors need and not all authors get. This book did not just get written out of verve and willpower.

Thank you to my parents, Jack Clancy and Janet Bridges-Clancy, to my aunt Bette Bridges, and to my sister Liz Lerner for the jokes, your belief in me, and so much more. Thank you also to the Yale Graduate Employees and Students Organization, now Local 33 UNITE HERE, for providing me with my most meaningful and important education.

Finally, thank you to my husband, Brendan Harley, and to my children, Sam and Edith. Bren, for over twenty years you have loved me exactly as I am—and I can be a real asshole. Knowing that you love me unconditionally has healed me and bolstered me in some of my most difficult moments. That you love my strength is what often keeps me strong. The ways you have grown over the years give me more hope about this world than just about anything else. I am so proud to know and love you.

Sammy and Edie, I just keep trying to do right by you and deserve you. I hope I get there. I love your curls; I love your smells; I love how you stand and walk and talk. Sam, you are one of the bravest yet most tender people I know. Edith, your imagination and creativity charm me to excess. I love you two more every day, and I delight in how you keep becoming more yourselves.

NOTES

Preface

1. Clancy, K. B. H., I. Nenko, and G. Jasienska, *Menstruation does not cause anemia: Endometrial thickness correlates positively with erythrocyte count and hemoglobin concentration in premenopausal women*, American Journal of Human Biology, 2006. **18**(5): 710–13.

2. Clancy, K. B. H., et al., *Survey of Academic Field Experiences (SAFE): Trainees report harassment and assault*, PLoS ONE, 2014. **9**(7): 1–9; Clancy, K. B. H., et al., *Double jeopardy in astronomy and planetary science: Women of color face greater risks of gendered and racial harassment*, JGR Planets, 2017. **122**(7): 1610–23; Nelson, R., et al., *Signaling safety: Characterizing fieldwork experiences and their implications for career trajectories*, American Anthropologist, 2017. **119**(4): 710–22; Aycock, L., et al., *Sexual harassment reported by undergraduate female physicists*, Physical Review Physics Education Research, 2019. **15**:010121; Richey, C. R., et al., *Gender and sexual minorities in astronomy and planetary science face increased risks of* harassment and assault, Bulletin of the American Astronomical Society, 2019. **51**(4), https://doi.org/10.3847/25c2cfeb.c985281e; Clancy, K. B. H., L. M. Cortina, and A. R. Kirkland, *Opinion: Use science to stop sexual harassment in higher education*, Proceedings of the National Academy of Sciences, 2020. **117**(37): 22614–18; Johnson, P., Widnall, S. E., and Benya, F. F., eds., *Sexual Harassment of Women: Climate, Culture, and Consequences in Academic Sciences, Engineering, and Medicine*. 2018: National Academies Press.

3. Baerwald, A. R., G. P. Adams, and R. A. Pierson, *Characterization of ovarian follicular wave dynamics in women*, Biology of Reproduction, 2003. **69**(3): 1023–31.

4. IJland, M. M., et al., *Endometrial wavelike movements during the menstrual cycle*, Fertility and Sterility, 1996. **65**(4): 746–49.

5. Insler, V., et al., *Sperm storage in the human cervix: A quantitative study*, Fertility and Sterility, 1980. **33**(3): 288–93.

6. Evans, J., et al., *Menstrual fluid factors facilitate tissue repair: Identification and functional action in endometrial and skin repair*, FASEB Journal, 2019. **33**(1):582–605.

Introduction: Taking the Mystery Out of Menstruation

1. Roberts, T. A., et al., *"Feminine protection": The effects of menstruation on attitudes toward women*, Psychology of Women Quarterly, 2002. **26**:131–39.

2. White, L. R., *The function of ethnicity, income level, and menstrual taboos in postmenarcheal adolescents' understanding of menarche and menstruation*, Sex Roles, 2013. **68**(1): 65–76.

3. Burrows, A., and S. Johnson, *Girls' experiences of menarche and menstruation*, Journal of Reproductive and Infant Psychology, 2005. **23**(3): 235–49.

4. McPherson, M. E., and L. Korfine, *Menstruation across time: Menarche, menstrual attitudes, experiences, and behaviors*, Women's Health Issues, 2004. **14**(6): 193–200.

5. Cheng, C. Y., K. Yang, and S. R. Liou, *Taiwanese adolescents' gender differences in knowledge and attitudes towards menstruation*, Nursing and Health Sciences, 2007. **9**(2): 127–34.

6. Brooks-Gunn, J., and D. N. Ruble, *Men's and women's attitudes and beliefs about the menstrual cycle*, Sex Roles, 1986. **14**(5–6): 287–99.

7. Wister, J. A., M. L. Stubbs, and C. Shipman, *Mentioning menstruation: A stereotype threat that diminishes cognition?*, Sex Roles, 2013. **68**(1): 19–31.

8. Klausen, S., and A. Bashford, *Fertility control: Eugenics, neo-Malthusianism, and feminism*, in *The Oxford Handbook of the History of Eugenics*. 2010: Oxford University Press, pp. 98–115.

9. Huff, C., *In Texas, Abortion Laws Inhibit Care for Miscarriages*, NPR, May 10, 2022.

10. Green, E., *State-Mandated Mourning for Aborted Fetuses*, Atlantic, May 14, 2016; Promislow, J. H., et al., *Bleeding following pregnancy loss before 6 weeks' gestation*, Human Reproduction, 2007. **22**(3): 853–57.

11. Owens, D. C., *Medical Bondage: Race, Gender, and the Origins of American Gynecology*. 2017: University of Georgia Press; Washington, H. A., *Medical Apartheid: The Dark History of Medical Experimentation on Black Americans from Colonial Times to the Present*. 2006: Doubleday Books.

12. Towghi, F., *Haunting expectations of hospital births challenged by traditional midwives*, Medical Anthropology, 2018. **37**(8): 674–87; Rebuelta-Cho, A. P., *"Give her the baby's hat so she can bite it": Obstetric violence in Flores, Indonesia*, Moussons-Recherche En Sciences Humaines Sur L'Asie Du Sud-Est, 2021. (38): 57–84; Bonaparte, A. D., *Physicians' discourse for establishing authoritative knowledge in birthing work and reducing the presence of the granny midwife*, Journal of Historical Sociology, 2015. **28**(2): 166–94; Theobald, B., *To instill the hospital habit*, in *Reproduction on the Reservation: Pregnancy, Childbirth, and Colonialism in the Long Twentieth Century*. 2019: University of North Carolina Press, pp. 44–70; Cheney, M. J., *Homebirth as systems-challenging praxis: Knowledge, power, and intimacy in the birthplace*, Qualitative Health Research, 2008. **18**(2): 254–67.

13. Singer, M. R., et al., *Pediatricians' knowledge, attitudes and practices surrounding menstruation and feminine products*, International Journal of Adolescent Medicine and Health, 2020. **34**(3): n.p.

14. Finn, C., *Menstruation: A nonadaptive consequence of uterine evolution*, Quarterly Review of Biology, 1998. **73**(2): 163–73.

15. Profet, M., *Menstruation as a defense against pathogens transported by sperm*, Quarterly Review of Biology, 1993. **68**(3): 335–86; Strassmann, B., *The evolution of endometrial cycles and menstruation*, Quarterly Review of Biology, 1996. **71**(2): 181–220.

16. Wrangham, R. W., and D. Peterson, *Demonic Males: Apes and the Origins of Human Violence*. 1996: Houghton Mifflin Harcourt; Schiebinger, L. *Primatology, archaeology, and human origins—feminist interventions*, in *Conference on Equal Rites, Unequal Outcomes—Women in American Research Universities*. 1998: Kluwer Academic/Plenum.

17. Rodrigues, M. A., *Emergence of sex-segregated behavior and association patterns in juvenile spider monkeys*, Neotropical Primates, 2014. **21**(2): 183–88; Rodrigues, M., et al., *From maternal*

tending to adolescent befriending: The adolescent transition of social support, American Journal of Primatology, 2019. **82**(11): e23050; Rodrigues, M. A., and E. R. Boeving, *Comparative social grooming networks in captive chimpanzees and bonobos*, Primates, 2019. **60**(3): 191–202; Lee, K. M. N., et al., *Physical activity in women of reproductive age in a transitioning rural Polish population*, American Journal of Human Biology, 2019. **31**(3): e23231; Sear, R., R. Mace, and I. A. McGregor, *Maternal grandmothers improve nutritional status and survival of children in rural Gambia*, Proceedings of the Royal Society of London Series B—Biological Sciences, 2000. **267**(1453): 1641–47; Bove, R. B., C. R. Valeggia, and P. T. Ellison, *Girl helpers and time allocation of nursing women among the Toba of Argentina*, Human Nature, 2002. **13**(4): 457–72; Schiebinger, L. *Primatology, archaeology, and human origins—feminist interventions. In Conference on Equal Rites, Unequal Outcomes—Women in American Research Universities*. 1998. Kluwer Academic/Plenum.

18. Moore, A. M., *Victorian medicine was not responsible for repressing the clitoris: Rethinking homology in the long history of women's genital anatomy*, Signs: Journal of Women in Culture and Society, 2018. **44**(1): 53–81.

19. Riddle, J. M., *Birth control (abortion and contraception)*, in *The Encyclopedia of Ancient History*. 2013: Wiley-Blackwell, pp. 1–2; Reiches, M., *Reproductive justice and the history of prenatal supplementation: Ethics, birth spacing, and the "priority infant" model in the Gambia: Winner of the 2019 Catharine Stimpson Prize for Outstanding Feminist Scholarship*, Signs: Journal of Women in Culture and Society, 2019. **45**(1): 3–26.

20. Strings, S., *The rise of the big Black woman*, in *Fearing the Black Body: The Racial Origins of Fat Phobia*. 2019: New York University Press.

21. Galton, F., *Inquiries into Human Faculty and Its Development*. 1883: Macmillan.

22. Martin, E., *The egg and the sperm: How science has constructed a romance based on stereotypical male-female roles*, Signs: Journal of Women in Culture and Society, 1991. **16**(3): 485–501.

23. Strings, S., *Birth of the ascetic aesthetic*, in *Fearing the Black Body: The Racial Origins of Fat Phobia*. 2019: New York University Press.

24. Shires, D. A., and K. Jaffee, *Factors associated with health care discrimination experiences among a national sample of female-to-male transgender individuals*, Health and Social Work, 2015. **40**(2): 134–41.

25. Besse, M., N. M. Lampe, and E. S. Mann, *Focus: Sex and reproduction: Experiences with achieving pregnancy and giving birth among transgender men: A narrative literature review*, Yale Journal of Biology and Medicine, 2020. **93**(4): 517; Dufour, D., *The energetic cost of physical activity and the regulation of reproduction*, in *Reproduction and Adaptation: Topics in Human Reproductive Ecology*, ed. C. Maschie-Taylor and L. Rosetta. 2011: Cambridge University Press; Nadeau, J. H., *Do gametes woo? Evidence for their nonrandom union at fertilization*, Genetics, 2017. **207**(2): 369–87; Firman, R. C., et al., *Postmating female control: 20 years of cryptic female choice*, Trends in Ecology and Evolution, 2017. **32**(5): 368–82; Insler, V., et al. *Sperm storage in the human cervix: A quantitative study*. Fertility and Sterility, 1980. **33**(3): 288–93.

26. Cryle, P., and E. Stephens, *Introduction*, in *Normality: A Critical Genealogy*. 2017: University of Chicago Press.

27. Insler, V., et al. *Sperm storage in the human cervix: A quantitative study*. Fertility and Sterility, 1980. **33**(3): 288–93.

28. Beisel, N., and T. Kay, *Abortion, race, and gender in nineteenth-century America*, American Sociological Review, 2004. **69**(4): 498–518.

29. Klausen, S., and A. Bashford. *Fertility control*. In *The Oxford Handbook of the History of Eugenics*, pp. 98–115.

30. Bryson, A., A. Koyama, and A. Hassan, *Addressing long-acting reversible contraception access, bias, and coercion: Supporting adolescent and young adult reproductive autonomy*, Current Opinion in Pediatrics, 2021. **33**(4): 345–53; Higgins, J. A., *Celebration meets caution: LARC's boons, potential busts, and the benefits of a reproductive justice approach*, Contraception, 2014. **89**(4): 237–41; Amico, J. R., et al., *"She just told me to leave it": Women's experiences discussing early elective IUD removal*, Contraception, 2016. **94**(4): 357–61; Amico, J. R., et al., *"$231 . . . to pull a string!!!" American IUD users' reasons for IUD self-removal: An analysis of internet forums*, Contraception, 2020. **101**(6): 393–98; Amico, J. R., et al., *Access to IUD removal: Data from a mystery-caller study*, Contraception, 2020. **101**(2): 122–29.

31. Murray, M., *Race-ing Roe: Reproductive justice, racial justice, and the battle for Roe v. Wade*, Harvard Law Review, 2020. **134**:2025.

32. Harrison, F. V., *The persistent power of "race" in the cultural and political economy of racism*, Annual Review of Anthropology, 1995: 47–74; Harrison, F. V., ed., *Decolonizing Anthropology: Moving Further toward an Anthropology for Liberation*. 1991: American Anthropological Association; Baker, L. D., *Anthropology and the Racial Politics of Culture*. 2010: Duke University Press; Brodkin, K., S. Morgen, and J. Hutchinson, *Anthropology as white public space?*, American Anthropologist, 2011. **113**(4): 545–56; Davis, J. L., and K. A. Smalls, *Dis/possession afoot: American (anthropological) traditions of anti-Blackness and coloniality*, Journal of Linguistic Anthropology, 2021. **31**(2): 275–82; Mullings, L., *Interrogating racism: Toward an antiracist anthropology*, Annual Review of Anthropology, 2005. **34**:667–93.

33. Tsing, A. L., *The Mushroom at the End of the World: On the Possibility of Life in Capitalist Ruins*. 2015: Princeton University Press.

34. Tsing, A. L., *The Mushroom at the End of the World: On the Possibility of Life in Capitalist Ruins*.

35. Shotwell, A., *Against Purity: Living Ethically in Compromised Times*. 2016: University of Minnesota Press.

36. Joel, D., *Genetic-gonadal-genitals sex (3G-sex) and the misconception of brain and gender, or, why 3G-males and 3G-females have intersex brain and intersex gender*, Biology of Sex Differences, 2012. **3**(1): 27; Barres, B., *Does gender matter?*, Nature, 2006. **442**:133–36; Steele, J. R., and N. Ambady, *"Math is hard!" The effect of gender priming on women's attitudes*, Journal of Experimental Social Psychology, 2006. **42**(4): 428–36; Prescod-Weinstein, C., *Ain't I a woman? At the intersection of gender, race, and sexuality*, Women In Astronomy (blog), 2014, http://womeninastronomy .blogspot.com/2014/05/aint-i-woman-at-intersection-of-gender.html; Pilar Matud, M., J. M. Bethencourt, and I. Ibáñez, *Relevance of gender roles in life satisfaction in adult people*, Personality and Individual Differences, 2014. **70**:206–11; Fausto-Sterling, A., *Sexing the Body: Gender Politics and the Construction of Sexuality*. 2000: Basic Books.

37. Kessler, S. J., *The medical construction of gender: Case management of intersexed infants*, Signs: Journal of Women in Culture and Society, 1990. **16**(1): 3–26.

38. Morrison, T., A. Dinno, and T. Salmon, *The erasure of intersex, transgender, nonbinary, and agender experiences through misuse of sex and gender in health research*, American Journal of

Epidemiology, 2021. **190**(12): 2712–17; Blackless, M., et al., *How sexually dimorphic are we? Review and synthesis*, American Journal of Human Biology, 2000. **12**(2): 151–66.

39. Fausto-Sterling, A., *Gender/sex, sexual orientation, and identity are in the body: How did they get there?*, Journal of Sex Research, 2019. **56**(4–5): 529–55; Morgenroth, T., and M. K. Ryan, *The effects of gender trouble: An integrative theoretical framework of the perpetuation and disruption of the gender/sex binary*, Perspectives on Psychological Science, 2020. **16**(6): 1113–42; DuBois, L. Z., et al., *Biocultural approaches to transgender and gender diverse experience and health: Integrating biomarkers and advancing gender/sex research*, American Journal of Human Biology, 2021. **33**(1): e23555.

40. Schuller, K., *The Biopolitics of Feeling*. 2018: Duke University Press.

41. Clancy, K., and J. Davis, *Soylent is people and WEIRD is white: Biological anthropology, whiteness, and the limits of the WEIRD*, Annual Review of Anthropology, 2019. **48**:169–86.

42. Liboiron, M., *Introduction*, in *Pollution Is Colonialism*. 2021, Duke University Press.

43. Shapin, S., *Never Pure: Historical Studies of Science as If It Was Produced by People with Bodies, Situated in Time, Space, Culture, and Society, and Struggling for Credibility and Authority*. 2010: Johns Hopkins University Press.

Chapter 1: There Is a Reason for All of This

1. Hong, J. S., and D. L. Espelage, *A review of research on bullying and peer victimization in school: An ecological system analysis*, Aggression and Violent Behavior, 2012. 17(4): 311–22; Vostral, S. L., *Under Wraps: A History of Menstrual Hygiene Technology*. 2008: Lexington Books.

2. Bajaj, M., A. Ghaffar-Kucher, and K. Desai, *Brown bodies and xenophobic bullying in US schools: Critical analysis and strategies for action*, Harvard Educational Review, 2016. 86(4): 481–505.

3. Willness, C. R., P. Steel, and K. Lee, *A meta-analysis of the antecedents and consequences of workplace sexual harassment*, Personnel Psychology, 2007. 60(1): 127–62; Berdahl, J. L., et al., *Work as a masculinity contest*, Journal of Social Issues, 2018. 74(3): 422–48.

4. Jones, K., and T. Okun, *White supremacy culture*, in *Dismantling Racism: A Workbook for Social Change Groups*. 2001: Dismantlingracism.org.

5. Landau, M., *Narratives of Human Evolution*. 1991: Yale University Press.

6. Strings, S., *Fearing the Black Body: The Racial Origins of Fat Phobia*. 2019: New York University Press; Dubow, S., *Human origins, race typology and the other Raymond Dart*, African Studies, 1996. 55(1): 1–30; Janik, E., *Marketplace of the Marvelous: The Strange Origins of Modern Medicine*. 2014: Beacon Press.

7. Shapin, S., *Placing the view from nowhere: Historical and sociological problems in the location of science*, Transactions of the Institute of British Geographers, 1998. 23(1): 5–12; Baker, K., M. P. Eichhorn, and M. Griffiths, *Decolonizing field ecology*, Biotropica, 2019. 51(3): 288–92; Nagel, T., *The View from Nowhere*. 1989: Oxford University Press.

8. Jacobs-Huey, L., *The natives are gazing and talking back: Reviewing the problematics of positionality, voice, and accountability among "native" anthropologists*, American Anthropologist, 2002. 104(3): 791–804; Clancy, K. B. H., and J. Davis. *Soylent is people and WEIRD is white: Biological anthropology, whiteness, and the limits of the WEIRD*. Annual Review of Anthropology, 2019. 48:169–86.

9. Buckley, T., and A. Gottlieb, eds., *Blood Magic: The Anthropology of Menstruation*. 1988, University of California Press: Berkeley.

10. *Merriam-Webster's Collegiate Dictionary*, vol. 2. 2004: Merriam-Webster.

11. Henare, M., *Tapu, mana, mauri, hau, wairua: A Maori philosophy of vitalism and cosmos*, Indigenous Traditions and Ecology, 2001: 197–221.

12. Sharma, B., and K. Schultz, *Woman and 2 Children Die in Nepal Menstruation Hut, New York Times*, January 9, 2019.

13. Amatya, P., et al., *Practice and lived experience of menstrual exiles (Chhaupadi) among adolescent girls in far-western Nepal*, PloS One, 2018. 13(12): e0208260.

14. Preiss, D., *A Young Woman Died in a Menstrual Hut in Nepal*, NPR, November 28, 2016; Sharma, B., and K. Schultz. *Woman and 2 Children Die in Nepal Menstruation Hut. New York Times*, January 9, 2019.

15. Buckley, T., *Menstruation and the Power of Yurok Women*, in Buckley and Gottlieb, *Blood Magic*.

16. Baldy, C. R., *We Are Dancing for You: Native Feminisms and the Revitalization of Women's Coming-of-Age Ceremonies*, Indigenous Confluences, ed. C. Cote and C. Thrush. 2018: University of Washington Press.

17. Buckley, T., and A. Gottlieb, *A critical appraisal of theories of menstrual symbolism*, in Buckley and Gottlieb, *Blood Magic*.

18. Gottlieb, A., *Rethinking female pollution: The Beng of Côte d'Ivoire*, Dialectical Anthropology, 1989. 14(2): 65–79.

19. Gottlieb, A., *Rethinking female pollution*.

20. Sniekers, M., *From little girl to young woman: The menarche ceremony in Fiji*, Fijian Studies, 2005. 3(2): 397.

21. Dharmalingam, A., *The implications of menarche and wedding ceremonies for the status of women in a South Indian village*, Indian Anthropologist, 1994. 24(1): 31–43.

22. Anderson, K., *Life Stages and Native Women: Memory, Teachings, and Story Medicine*, Critical Studies in Native History, ed. J. Brownlie. 2011: University of Manitoba Press.

23. Amos, V., and P. Parmar, *Challenging imperial feminism*, Feminist Review, 1984. 17(1): 3–19.

24. Buckley, T., *Menstruation and the power of Yurok women: Methods in cultural reconstruction*, American Ethnologist, 1982. 9(1): 47–60.

25. Baldy, C. R. *We Are Dancing for You*.

26. Rodnite Lemay, H., *Womens Secrets: A Translation of Pseudo-Albertus Magnus' de Secretis Mulierum with Commentaries*. 1992: State University of New York Press.

27. Bildhauer, B., *Medieval Blood*. 2006: University of Wales Press.

28. Hindson, B., *Attitudes towards menstruation and menstrual blood in Elizabethan England*, Journal of Social History, 2009: 89–114.

29. Nissen, A., *Transgression, pollution, deformity, bewitchment: Menstruation and supernatural threat in late medieval and early modern England, 1250–1750*. PhD diss., State University of New York Empire State College, 2017.

30. Healy, M., and W. Tagoe, *Dangerous blood: Menstruation, medicine and myth in early modern England*, in *National Healths: Gender, Sexuality and Health in a Cross-Cultural Context*, ed. M. Wooten and W. Tagoe, pp. 83–95. 2013: Routledge.

31. Macht, D. I., and D. S. Lubin, *A phyto-pharmacological study of menstrual toxin*, Journal of Pharmacology and Experimental Therapeutics, 1923. 22(5): 413–66.

32. Pickles, V., *Prostaglandins and dysmenorrhea: Historical survey*, Acta Obstetricia et Gynecologica Scandinavica suppl., 1979. 87:7–12.

33. Perlstein, M., and A. Matheson, *Allergy due to menotoxin of pregnancy*, Archives of Pediatrics and Adolescent Medicine, 1936. 52(2): 303.

34. Ashley-Montagu, M., *Physiology and the origins of the menstrual prohibitions*, Quarterly Review of Biology, 1940. 15(2): 211–20.

35. Reid, H., *Letter: The brass-ring sign*, Lancet, 1974. 1(7864): 988.

36. Profet, M. *Menstruation as a defense against pathogens transported by sperm.* Quarterly Review of Biology, 1993. 68(3): 335–86.

37. Strassmann, B. I. *The evolution of endometrial cycles and menstruation.* Quarterly Review of Biology, 1996. 71(2): 181–220.

38. Strassmann, B., *Energy economy in the evolution of menstruation*, Evolutionary Anthropology, 1996. 5(5): 157–64; Strassmann, B. I. *The evolution of endometrial cycles and menstruation.*

39. Serin, I. S., et al., *Effects of hypertension and obesity on endometrial thickness*, European Journal of Obstetrics and Gynaecology and Reproductive Biology, 2003. 109:72–75.

40. Clancy, K. B. H., *Unexpected luteal endometrial decline in a healthy rural Polish population*, European Journal of Obstetrics and Gynaecology and Reproductive Biology, 2007. 134(1): 133–34; Clancy, K. B. H., et al., *Endometrial thickness is not independent of luteal phase day in a rural Polish population*, Anthropological Science, 2009. 117(3): 157–63.

41. Pontzer, H., et al., *Energy expenditure and activity among Hadza hunter-gatherers*, American Journal of Human Biology, 2015. 27(5): 628–37.

42. Bentley, G. R., A. M. Harrigan, and P. T. Ellison, *Ovarian cycle length and days of menstruation of Lese horticulturalists (abstract)*, American Journal of Physical Anthropology, 1990. 81:193–94.

43. Strassmann, B. I., *The biology of menstruation in Homo sapiens: Total lifetime menses, fecundity, and nonsynchrony in a natural-fertility population*, Current Anthropology, 1997. 38(1): 123–29.

44. Prior, J. C., et al., *Ovulation prevalence in women with spontaneous normal-length menstrual cycles—a population-based cohort from HUNT3, Norway*, PLOS ONE, 2015. 10(8): e0134473.

45. Brosens, I., et al., *The impact of uterine immaturity on obstetrical syndromes during adolescence*, American Journal of Obstetrics and Gynecology, 2017. 217(5): 546–55; Brosens, J. J., et al., *A role for menstruation in preconditioning the uterus for successful pregnancy*, American Journal of Obstetrics and Gynecology, 2009. 200(6): 615.e1–.e6.

46. Jabbour, H., et al., *Endocrine regulation of menstruation*, Endocrine Reviews, 2006. 27(1): 17–46.

47. Bianco, P., P. Gehron Robey, and P. Simmons, *Mesenchymal stem cells: Revisiting history, concepts, and assays*, Cell Stem Cell, 2008. 2(4): 313–19.

48. Evans, J., et al. *Menstrual fluid factors facilitate tissue repair: Identification and functional action in endometrial and skin repair.* FASEB Journal, 2019. 33:000.

49. Abalos, E., et al., *Global and regional estimates of preeclampsia and eclampsia: A systematic review*, European Journal of Obstetrics and Gynecology and Reproductive Biology, 2013. 170(1): 1–7.

50. Urquia, M., et al., *Disparities in pre-eclampsia and eclampsia among immigrant women giving birth in six industrialised countries*, BJOG: An International Journal of Obstetrics and Gynaecology, 2014. 121(12): 1492–500.

51. Dressler, W. W., K. S. Oths, and C. C. Gravlee, *Race and ethnicity in public health research: Models to explain health disparities*, Annual Review of Anthropology, 2005. 34:231–52.

52. Farley, K. E., et al., *The association between contraceptive use at the time of conception and hypertensive disorders during pregnancy: A retrospective cohort study of PRAMS participants*, Maternal and Child Health Journal, 2014. 18(8): 1779–85.

53. Vitzthum, V. J., et al., *Vaginal bleeding patterns among rural highland Bolivian women: Relationship to fecundity and fetal loss*, Contraception, 2001. 64:319–25.

54. Rogers-LaVanne, M., and K. Clancy, *Menstruation: Causes, consequences, and context*, in *Routledge Handbook of Anthropology and Reproduction*, ed. S. Han and C. Tomori. 2021: Routledge.

Chapter 2: Norma's "Normal" Cycle

1. Shapiro, H. L., *A portrait of the American people*, Natural History, 1945. **54**:248–55.

2. Stephens, E., *The Object of Normality: The "Search for Norma" Competition*, Queer Objects Symposium, Gender Institute and School of Literature, Languages and Linguistics, Australian National University, Canberra, 2014.

3. Novak, M., *Reckless Breeding of the Unfit: Earnest Hooton, Eugenics and the Human Body of the Year 2000*, *Smithsonian* magazine, February 12, 2013, https://www.smithsonianmag.com/history/reckless-breeding-of-the-unfit-earnest-hooton-eugenics-and-the-human-body-of-the-year-2000-15933294/.

4. Rose, T., *When U.S. Air Force Discovered the Flaw of Averages*, Star (Toronto), January 16, 2016.

5. Houghton, L., et al., *A migrant study of pubertal timing and tempo in British-Bangladeshi girls at varying risk for breast cancer*, Breast Cancer Research, 2014. 16:469.

6. Gray, S. H., et al., *Salivary progesterone levels before menarche: A prospective study of adolescent girls*, Journal of Clinical Endocrinology and Metabolism, 2010. 95(7): 3507–11.

7. Elson, J., *Menarche, menstruation, and gender identity: Retrospective accounts from women who have undergone premenopausal hysterectomy*, Sex Roles, 2002. 46(1–2): 37–48.

8. Cozzi, D., and V. Vinel, *Risky, early, controversial: Puberty in medical discourses*, Social Science & Medicine, 2015. 143:287–96.

9. Kramer, K. L., *Early sexual maturity among Pume foragers of Venezuela: Fitness implications of teen motherhood*, American Journal of Physical Anthropology, 2008. 136(3): 338–50.

10. Clancy, K. B. H., et al., *Relationships between biomarkers of inflammation, ovarian steroids, and age at menarche in a rural Polish sample*, American Journal of Human Biology, 2013. 25(3): 389–98.

11. Rogers, M. P., et al., *Declining ages at menarche in an agrarian rural region of Poland*, American Journal of Human Biology, 2020. 32(3): e23362.

12. Rogers, M., et al., *The effects of adolescent-parent open communication on age at menarche*. Forthcoming; Walker, R., et al., *Growth rates and life histories in twenty-two small-scale societies*, American Journal of Human Biology, 2006. 18(3): 295–311.

13. Sear, R., P. Sheppard, and D. A. Coall, *Cross-cultural evidence does not support universal acceleration of puberty in father-absent households*, Philosophical Transactions of the Royal Society B: Biological Sciences, 2019. 374(1770): 20180124.

14. Rodrigues, M. A., *Stress and sociality in a patrilocal primate: Do female spider monkeys tend-and-befriend?* PhD diss., Ohio State University, 2013; Rodrigues, M. A. *Emergence of sex-segregated behavior and association patterns in juvenile spider monkeys.* Neotropical Primates, 2014. **21**(2): 183–88; Rodrigues, M. A., and E. R. Boeving. *Comparative social grooming networks in captive chimpanzees and bonobos.* Primates, 2019. 60(3): 191–202.

15. Almeling, R., *GUYnecology: The Missing Science of Men's Reproductive Health.* 2020: University of California Press.

16. Almeling, R., *Sex, sperm, and fatherhood*, in Almeling, *GUYnecology*, chap. 5.

17. Murdock, M., *The Heroine's Journey: Woman's Quest for Wholeness.* 30th anniversary ed. 2020: Shambhala.

18. Forabosco, A., et al., *Morphometric study of the human neonatal ovary*, Anatomical Record, 1991. 231(2): 201–8; Albamonte, M. I., et al., *The infant and pubertal human ovary: Balbiani's body-associated VASA expression, immunohistochemical detection of apoptosis-related BCL2 and BAX proteins, and DNA fragmentation*, Human Reproduction, 2013. 28(3): 698–706.

19. Vanden Brink, H., et al., *Age-related changes in major ovarian follicular wave dynamics during the human menstrual cycle*, Menopause, 2013. 20(12): 1243–54; Baerwald, A., et al., *Age-related changes in luteal dynamics: Preliminary associations with antral follicular dynamics and hormone production during the human menstrual cycle*, Menopause, 2018. 25(4): 399–407.

20. Baerwald, A. R., O. Olatunbosun, and R. A. Pierson, *Effects of oral contraceptives administered at defined stages of ovarian follicular development*, Fertility and Sterility, 2006. 86(1): 27–35; Baerwald, A. R., O. A. Olatunbosun, and R. A. Pierson, *Ovarian follicular development is initiated during the hormone-free interval of oral contraceptive use*, Contraception, 2004. 70(5): 371–77.

21. Robertson, D. M., et al., *Random start or emergency IVF/in vitro maturation: A new rapid approach to fertility preservation*, Women's Health, 2016. 12(3): 339–49.

22. Mihm, M., and A. Evans, *Mechanisms for dominant follicle selection in monovulatory species: A comparison of morphological, endocrine and intraovarian events in cows, mares and women*, Reproduction in Domestic Animals, 2008. 43:48–56.

23. Prior, J. C., and C. L. Hitchcock, *The endocrinology of perimenopause: Need for a paradigm shift*, Frontiers in Bioscience, 2011. 3:474–86.

24. Prior, J. C., *Perimenopause: The complex endocrinology of the menopausal transition*, Endocrine Reviews, 1998. 19(4): 397–428.

25. Park, S. U., L. Walsh, and K. M. Berkowitz, *Mechanisms of ovarian aging*, Reproduction, 2021. 162(2): R19–R33.

26. Sievert, L. L., *Menopause across cultures: Clinical considerations*, Menopause, 2014. 21(4): 421–23.

27. Gottschalk, M. S., et al., *The relation of number of childbirths with age at natural menopause: A population study of 310 147 women in Norway*, Human Reproduction, 2022. 37(2): 333–40.

28. Schoenaker, D. A., et al., *Socioeconomic position, lifestyle factors and age at natural menopause: A systematic review and meta-analyses of studies across six continents*, International Journal of Epidemiology, 2014. 43(5): 1542–62.

29. Ghaderi, D., et al., *Sexual selection by female immunity against paternal antigens can fix loss of function alleles*, Proceedings of the National Academy of Sciences, 2011. 108(43): 17443–48; Nadeau, J. H. *Do gametes woo? Evidence for their nonrandom union at fertilization*. Genetics, 2017. 207(2): 369–87; Firman, R. C., et al. *Postmating female control: 20 years of cryptic female choice*. Trends in Ecology and Evolution, 2017. 32(5): 368–82.

30. Parker, R., T. Larkin, and J. Cockburn, *A visual analysis of gender bias in contemporary anatomy textbooks*, Social Science and Medicine, 2017. 180:106–13.

31. Nixon, B., R. J. Aitken, and E. A. McLaughlin, *New insights into the molecular mechanisms of sperm-egg interaction*, Cellular and Molecular Life Sciences, 2007. 64(14): 1805–23.

32. Wilcox, A. J., D. D. Baird, and C. R. Weinberg, *Time of implantation of the conceptus and loss of pregnancy*, New England Journal of Medicine, 1999. 340(23): 1796–99.

33. IJland, M. M., et al., *Relation between endometrial wavelike activity and fecundability in spontaneous cycles*, Fertility and Sterility, 1997. 67(3): 492–96.

34. Gupta, S. K., *The human egg's zona pellucida*, Current Topics in Developmental Biology, 2018. 130:379–411.

35. Ellington, J. E., et al., *Higher-quality human sperm in a sample selectively attach to oviduct (fallopian tube) epithelial cells in vitro*, Fertility and Sterility, 1999. 71(5): 924–29.

36. Wang, H., and S. K. Dey, *Roadmap to embryo implantation: Clues from mouse models*, Nature Reviews Genetics, 2006. 7(3): 185–99; Maia, J., et al., *The endocannabinoid system expression in the female reproductive tract is modulated by estrogen*, Journal of Steroid Biochemistry and Molecular Biology, 2017. 174:40–47; Taylor, A. H., et al., *Histomorphometric evaluation of cannabinoid receptor and anandamide modulating enzyme expression in the human endometrium through the menstrual cycle*, Histochemistry and Cell Biology, 2010. 133(5): 557–65.

37. O'Rourke, M. T., and P. T. Ellison, *Age-related patterns of salivary estradiol across the menstrual-cycle*, American Journal of Physical Anthropology, 1989. 78(2): 281.

Chapter 3: The Stress of Energy

1. Clarke, E. H., *Sex in Education; or, A Fair Chance for the Girls*. 1873: James R. Osgood.

2. Oreskes, N., *Science awry*. In *Why Trust Science?*, ed. N. Oreskes. 2019: Princeton University Press, pp. 69–146; Seller, M. S., *Dr. Clarke vs. "the Ladies": Coeducation and Women's Roles in the 1870's*. Paper presented at the Annual Meeting of the American Educational Research Association, Montreal, 1983.

3. Oreskes, N. *Science awry*.

4. Daggett, C. N., *The Birth of Energy: Fossil Fuels, Thermodynamics, and the Politics of Work*, Elements, ed. S. Alaimo and N. Starosielski. 2019: Duke University Press.

5. Subar, A. F., et al., *Addressing current criticism regarding the value of self-report dietary data*, Journal of Nutrition, 2015. 145(12): 2639–45; Slootmaker, S. M., et al., *Disagreement in physical activity assessed by accelerometer and self-report in subgroups of age, gender, education and weight status*, International Journal of Behavioral Nutrition and Physical Activity, 2009. 6(1): 1–10.

6. Black, P., and U. Sharma, *Men are real, women are "made up": Beauty therapy and the construction of femininity*, Sociological Review, 2001. 49(1): 100–116.

7. Hrdy, S. B., *Mothers and others*, Natural History, 2001. 110(4): 50–63.

8. Thyfault, J. P., et al., *Physiology of sedentary behavior and its relationship to health outcomes*, Medicine and Science in Sports and Exercise, 2015. 47(6): 1301; Patterson, R., et al., *Sedentary behaviour and risk of all-cause, cardiovascular and cancer mortality, and incident type 2 diabetes: A systematic review and dose response meta-analysis*, European Journal of Epidemiology, 2018. 33(9): 811–29.

9. Ortega, F. B., et al., *The Fat but Fit Paradox: What We Know and Don't Know about It*. 2018: BMJ and British Association of Sport and Exercise Medicine.

10. *In Memoriam: Rose Epstein Frisch, Expert in Women's Fertility*, T. H. Chan School of Public Health, Harvard University, 2015, https://www.hsph.harvard.edu/news/features/in-memoriam -rose-epstein-frisch-expert-in-womens-fertility-2/.

11. Crawford, J. D., and D. C. Osler, *Body composition at menarche: The Frisch-Revelle hypothesis revisited*, Pediatrics, 1975. 56(3): 449–58; Trussell, J., *Menarche and fatness: Reexamination of the critical body composition hypothesis*, Science, 1978. 200:1506–9; Trussell, J., *Statistical flaws in evidence for the Frisch hypothesis that fatness triggers menarche*, Human Biology, 1980. 52(4): 711.

12. Trussell, J. *Statistical flaws in evidence for the Frisch hypothesis that fatness triggers menarche*. Human Biology, 1980. 52(4): 711.

13. Frisch, R. E., and J. W. McArthur, *Menstrual cycles: Fatness as a determinant of minimum weight for height necessary for their maintenance or onset*, Science, 1974. 185(4155): 949–51.

14. Ellison, P. T., *Skeletal growth, fatness, and menarcheal age—a comparison of 2 hypotheses*, Human Biology, 1982. 54(2): 269–81.

15. Ellison, P. T., *On Fertile Ground*. 2001: Harvard University Press.

16. Oreskes, N., *Why trust science? Perspectives from the history and philosophy of science*, in Oreskes, *Why Trust Science?*, pp. 15–68.

17. Norgan, N., *The beneficial effects of body fat and adipose tissue in humans*, International Journal of Obesity, 1997. 21(9): 738–46.

18. Bekker, S., *On the History of (the Segregation of) Women's Sport, Holding Space*, March 20, 2022.

19. *Lady with Desire to Run Crashed Marathon, New York Times*, April 23, 1967.

20. Broocks, A., et al., *Cyclic ovarian function in recreational athletes*, Journal of Applied Physiology, 1990. 68(5): 2083–86; Wojtys, E. M., et al., *Athletic activity and hormone concentrations in high school female athletes*, Journal of Athletic Training, 2015. 50(2): 185–92.

21. Ellison, P. T., et al., *The ecological context of human ovarian function*, Human Reproduction, 1993. 8(12): 2248–58; Jasienska, G., and P. T. Ellison, *Physical work causes suppression of ovarian function in women*, Proceedings of the Royal Society of London—Series B, 1998. 265(1408): 1847–51; Reed, J. L., et al., *Energy availability discriminates clinical menstrual status in exercising women*, Journal of the International Society of Sports Nutrition, 2015. 12(1): 11; Williams, N. I., S. M. Statuta, and A. Austin, *Female athlete triad: Future directions for energy availability and eating disorder research and practice*, Clinics in Sports Medicine, 2017. 36(4): 671–86; Huhmann, K., *Menses requires energy: A review of how disordered eating, excessive exercise, and high stress lead to menstrual irregularities*, Clinical Therapeutics, 2020. 42(3): 401–7.

22. Sokoloff, N. C., M. Misra, and K. E. Ackerman, *Exercise, training, and the hypothalamic-pituitary-gonadal axis in men and women*, Sports Endocrinology, 2016. 47:27–43.

23. Walberg-Rankin, J., W. Franke, and F. Gwazdauskas, *Response of beta-endorphin and estradiol to resistance exercise in females during energy balance and energy restriction*, International Journal of Sports Medicine, 1992. 13(7): 542–47.

24. Jasienska, G., and P. T. Ellison, *Energetic factors and seasonal changes in ovarian function in women from rural Poland*, American Journal of Human Biology, 2004. 16:563–80; Jasienska, G., et al., *Habitual physical activity and estradiol levels in women of reproductive age*, European Journal of Cancer Prevention, 2006. 15:439–45; Jasienska, G., and P. T. Ellison. *Physical work causes suppression of ovarian function in women*. Proceedings of the Royal Society of London—Series B, 1998. 265(1408): 1847–51.

25. Lee, K., et al., *Bone density and frame size in adult women: Effects of body size, habitual use, and life history*. Submitted.

26. Jasienska, G., and I. Thune, *Lifestyle, hormones, and risk of breast cancer*, British Medical Journal, 2001. 322:586–87; Jasienska, G., *The Fragile Wisdom: An Evolutionary View on Women's Biology and Health*. 2013: Harvard University Press; Lee, K., et al., *Bone density and frame size in adult women: Effects of body size, habitual use, and life history*, American Journal of Human Biology, 2020. 33(2): e23502; Jasienska, G., and P. T. Ellison. *Physical work causes suppression of ovarian function in women*. Proceedings of the Royal Society of London—Series B, 1998. 265(1408): 1847–51.

27. Bentley, G. R., et al., *Women's strategies to alleviate nutritional stress in a rural African society*, Social Science and Medicine, 1999. 48(2): 149–62; Pontzer, H., et al. *Energy expenditure and activity among Hadza hunter-gatherers*. American Journal of Human Biology, 2015. 27(5): 628–37.

28. Steinfeldt, J. A., et al., *Muscularity beliefs of female college student-athletes*, Sex Roles, 2011. 64(7–8): 543–54.

29. Prior, J., *Luteal phase defects and anovulation: Adaptive alterations occurring with conditioning exercise*, Seminars in Reproductive Endocrinology, 1985. 3:27–33.

30. Prior, J., et al., *Reversible luteal phase changes and infertility associated with marathon training*, Lancet, 1982. 320(8292): 269–70.

31. De Souza, M. J., and N. I. Williams, *Physiological aspects and clinical sequelae of energy deficiency and hypoestrogenism in exercising women*, Human Reproduction Update, 2004. 10(5): 433–48.

32. Mather, V., *Cain's Abuse Allegations against Salazar Cause More Upheaval in Track World*, New York Times, November 8, 2019; Futterman, M., *Another of Alberto Salazar's Runners Says He Ridiculed Her Body for Years*, New York Times, November 14, 2019; Cain, M., *I Was the Fastest Girl in America, until I Joined Nike*, New York Times, November 7, 2019.

33. Aschwanden, C., *Good to Go: What the Athlete in All of Us Can Learn from the Strange Science of Recovery*. 2019: W. W. Norton.

34. Strout, E., *That's Not Fat: How Ryan Hall Gained 40 Pounds of Muscle*, Runner's World, May 3, 2016, https://www.runnersworld.com/news/a20794927/thats-not-fat-how-ryan-hall-gained-40-pounds-of-muscle/.

35. Brant, J., *Duel in the Sun*, Runner's World, April 4, 2004, https://www.runnersworld.com/runners-stories/a21751476/duel-in-the-sun-salazar-beardsley/; Aschwanden, C. *Good to Go: What the Athlete in All of Us Can Learn from the Strange Science of Recovery*. 2019: W. W. Norton.

36. Mountjoy, M., et al., *The IOC consensus statement: Beyond the female athlete triad—relative energy deficiency in sport (RED-S)*, British Journal of Sports Medicine, 2014. 48(7): 491–97.

37. Brewis, A. A., *Stigma and the perpetuation of obesity*, Social Science and Medicine, 2014. 118:152–58; Lee, J. A., and C. J. Pausé, *Stigma in practice: Barriers to health for fat women*, Frontiers in Psychology, 2016. 7:2063; Tomiyama, A. J., et al., *How and why weight stigma drives the obesity "epidemic" and harms health*, BMC Medicine, 2018. 16(1): 123.

38. Jia, M., K. Dahlman-Wright, and J.-Å. Gustafsson, *Estrogen receptor alpha and beta in health and disease*, Best Practice and Research Clinical Endocrinology and Metabolism, 2015. 29(4): 557–68.

39. Ziomkiewicz, A., et al., *Body fat, energy balance and estradiol levels: A study based on hormonal profiles from complete menstrual cycles*, Human Reproduction, 2008. 23(11): 2555–63.

40. Yeung, E. H., et al., *Adiposity and sex hormones across the menstrual cycle: The BioCycle study*, International Journal of Obesity, 2013. 37(2): 237.

41. Major, B., A. J. Tomiyama, and J. M. Hunger, *The Negative and Bidirectional Effects of Weight Stigma on Health*. 2018: Oxford Handbooks Online. https://www.dishlab.org/pubs/Negative%20and%20Bidirectional%20Effects%20of%20Weight%20Stigma%20on%20Health%20-%20Oxford%20Handbooks.pdf.; Lee, J. A., and C. J. Pausé. *Stigma in practice: Barriers to health for fat women*. Frontiers in Psychology, 2016. 7:2063.

42. Green, B. B., N. S. Weiss, and J. R. Daling, *Risk of ovulatory infertility in relation to body weight*, Fertility and Sterility, 1988. 50(5): 721–26.

43. Tehrani, F. R., et al., *The prevalence of polycystic ovary syndrome in a community sample of Iranian population: Iranian PCOS prevalence study*, Reproductive Biology and Endocrinology, 2011. 9(1): 39; Norman, R., S. Mahabeer, and S. Masters, *Ethnic differences in insulin and glucose response to glucose between white and Indian women with polycystic ovary syndrome*, Fertility and Sterility, 1995. 63(1): 58–62.

44. Chen, X., et al., *Prevalence of polycystic ovary syndrome in unselected women from southern China*, European Journal of Obstetrics and Gynecology and Reproductive Biology, 2008. 139(1): 59–64.

45. Legro, R. S., *Effects of obesity treatment on female reproduction: Results do not match expectations*, Fertility and Sterility, 2017. 107(4): 860–67.

Chapter 4: Inflamed Cycles

1. Clancy, K., Twitter post, 2018, https://twitter.com/KateClancy/status/948788429860605952.

2. Fuller, E. A., et al., *Neuroimmune regulation of female reproduction in health and disease*, Current Opinion in Behavioral Sciences, 2019. 28:8–13; Beeler, J., F. Varricchio, and R. Wise, *Thrombocytopenia after immunization with measles vaccines: Review of the vaccine adverse events reporting system (1990 to 1994)*, Pediatric Infectious Disease Journal, 1996. 15(1): 88–90; Tarawneh, O., and H. Tarawneh, *Immune thrombocytopenia in a 22-year-old post Covid-19 vaccine*, American Journal of Hematology, 2021. 96(5): E133–34.

3. Thometz, K., *Do COVID-19 Vaccines Impact Menstrual Cycles?*, WTTW News (Chicago), May 7, 2021; Villareal, A., *"No Data" Linking Covid Vaccines to Menstrual Changes, US Experts Say*, Guardian (United States), April 23, 2021; Caron, C., *What Women Need to Know about the Covid Vaccine*, New York Times, September 13, 2021; Stock, N., *Some People Are Reporting*

Abnormal Periods after a COVID-19 Vaccine. U. of I. Professor Is Looking for Answers, Chicago Tribune, April 20, 2021; Lilly, S., *Can COVID-19 Vaccine Impact Your Menstrual Cycle? Doctors Address Side-Effects Concerns. Dr. Viray: "That's Just Not How These Vaccines Work,"* 6 News Richmond, April 14, 2021; Williams, C., *"No Concerns That We're Seeing": Utah Doctors Explain How COVID-19 Vaccines Factor in Women's Health, Pregnancy*, KSL.com (Salt Lake City), April 16, 2021; Cuda, A., *Is There a Link between COVID Vaccine and "Funky" Menstrual Periods? Experts Say It's Too Soon to Know*, Connecticut Post, April 19, 2021.

4. Pascal, M., et al., *Microbiome and allergic diseases*, Frontiers in Immunology, 2018. 9:1584.

5. Guarner, F., et al., *Mechanisms of disease: The hygiene hypothesis revisited*, Nature Clinical Practice Gastroenterology and Hepatology, 2006. 3(5): 275–84; Bloomfield, S. F., et al., *Time to abandon the hygiene hypothesis: New perspectives on allergic disease, the human microbiome, infectious disease prevention and the role of targeted hygiene*, Perspectives in Public Health, 2016. 136(4): 213–24.

6. Hyde, J. S., *The gender similarities hypothesis*, American Psychologist, 2005. 60(6): 581.

7. Fairweather, D., S. Frisancho-Kiss, and N. R. Rose, *Sex differences in autoimmune disease from a pathological perspective*, American Journal of Pathology, 2008. 173(3): 600–609.

8. Shah, R. and D. C. Newcomb, *Sex bias in asthma prevalence and pathogenesis*, Frontiers in Immunology, 2018. 9:2997.

9. Weinand, J. D., and J. D. Safer, *Hormone therapy in transgender adults is safe with provider supervision: A review of hormone therapy sequelae for transgender individuals*, Journal of Clinical and Translational Endocrinology, 2015. 2(2): 55–60.

10. Heuertz, R. M., et al., *Native and modified C-reactive protein bind different receptors on human neutrophils*, International Journal of Biochemistry and Cell Biology, 2005. 37(2): 320–35; Verma, S., P. E. Szmitko, and E.T.H. Yeh, *C-reactive protein: Structure affects function*, Circulation, 2004. 109(16): 1914–17.

11. Sproston, N. R., and J. J. Ashworth, *Role of C-reactive protein at sites of inflammation and infection*, Frontiers in Immunology, 2018. 9:754.

12. Petersen, A. M. W., and B. K. Pedersen, *The anti-inflammatory effect of exercise*, Journal of Applied Physiology, 2005. 98(4): 1154–62.

13. Clancy, K., A. Baerwald, and R. Pierson, *Systemic inflammation is associated with ovarian follicular dynamics during the human menstrual cycle*, PLoS One, 2013. 8(5): e64807.

14. Maybin, J. A., and H. O. Critchley, *Menstrual physiology: Implications for endometrial pathology and beyond*, Human Reproduction Update, 2015. 21(6): 748–61.

15. Wunder, D. M., et al., *Serum leptin and C-reactive protein levels in the physiological spontaneous menstrual cycle in reproductive age women*, European Journal of Endocrinology, 2006. 155(1): 137–42; Wander, K., E. Brindle, and K. A. O'Connor, *C-reactive protein across the menstrual cycle*, American Journal of Physical Anthropology, 2008. 136(2): 138–46; Blum, C., et al., *Low-grade inflammation and estimates of insulin resistance during the menstrual cycle in lean and overweight women*, Journal of Clinical Endocrinology and Metabolism, 2005. 90(6): 3230–35.

16. Bobel, C., *The Managed Body: Developing Girls and Menstrual Health in the Global South*. 2019: Palgrave Macmillan.

17. Bobel, C. *The Managed Body*.

18. Vostral, S. L. *Under Wraps: A History of Menstrual Hygiene Technology*. 2008: Lexington Books.

19. Martin, E., *The Woman in the Body: A Cultural Analysis of Reproduction*. 1980: Beacon Press.

20. Liboiron, M., *Pollution Is Colonialism*. 2021: Duke University Press.

21. Hennegan, J., and P. Montgomery, *Do menstrual hygiene management interventions improve education and psychosocial outcomes for women and girls in low and middle income countries? A systematic review*, PLOS ONE, 2016. **11**(2): e0146985; Phillips-Howard, P. A., et al., *Menstrual cups and sanitary pads to reduce school attrition, and sexually transmitted and reproductive tract infections: A cluster randomised controlled feasibility study in rural western Kenya*, BMJ Open, 2016. 6(11): e013229.

22. Ahmed, S., and A. Avasarala, *Urinary tract infections (UTI) among adolescent girls in rural Karimnagar District, AP KAP study*, Indian Journal of Preventive and Social Medicine, 2008. 39(1): 6–9.

23. Torondel, B., et al., *Association between unhygienic menstrual management practices and prevalence of lower reproductive tract infections: A hospital-based cross-sectional study in Odisha, India*, BMC Infectious Diseases, 2018. 18(1): 1–12.

24. Janoowalla, H., et al., *The impact of menstrual hygiene management on adolescent health: The effect of Go! pads on rate of urinary tract infection in adolescent females in Kibogora, Rwanda*, International Journal of Gynecology and Obstetrics, 2020. 148(1): 87–95.

25. Wamoyi, J., et al., *Transactional sex amongst young people in rural northern Tanzania: An ethnography of young women's motivations and negotiation*, Reproductive Health, 2010. 7(1): 2; Oruko, K., et al., *"He is the one who is providing you with everything so whatever he says is what you do": A qualitative study on factors affecting secondary schoolgirls' dropout in rural Western Kenya*, PLOS ONE, 2015. 10(12): e0144321.

26. Liboiron, M. *Pollution Is Colonialism*. 2021.

27. Abimbola, S., *The foreign gaze: Authorship in academic global health*, BMJ Specialist Journals, 2019: e002068.

28. Roved, J., H. Westerdahl, and D. Hasselquist, *Sex differences in immune responses: Hormonal effects, antagonistic selection, and evolutionary consequences*, Hormones and Behavior, 2017. 88:95–105.

29. Mann, E. S., *The power of persuasion: Normative accountability and clinicians' practices of contraceptive counseling*, SSM-Qualitative Research in Health, 2022. 2:100049; Odwe, G., et al., *Method-specific beliefs and subsequent contraceptive method choice: Results from a longitudinal study in urban and rural Kenya*, PLOS ONE, 2021. 16(6): e0252977; Fielding-Singh, P., and A. Dmowska, *Obstetric gaslighting and the denial of mothers' realities*, Social Science and Medicine, 2022: 114938; Werner, A., L. W. Isaksen, and K. Malterud, *"I am not the kind of woman who complains of everything": Illness stories on self and shame in women with chronic pain*, Social Science and Medicine, 2004. 59(5): 1035–45; Sebring, J. C., *Towards a sociological understanding of medical gaslighting in western health care*, Sociology of Health and Illness, 2021. 43(9): 1951–64; Higgins, J. A. Celebration meets caution: LARC's boons, potential busts, and the benefits of a reproductive justice approach. Contraception, 2014. 89(4): 237–41.

30. Male, V., *Are COVID-19 vaccines safe in pregnancy?*, Nature Reviews Immunology, 2021. 21(4): 200–201; Watson, R. E., T. B. Nelson, and A. L. Hsu, *Fertility considerations: The*

COVID-19 disease may have a more negative impact than the COVID-19 vaccine, especially among men, Fertility and Sterility, March 19, 2021, https://www.fertstertdialog.com/posts/fertility-considerations-the-covid-19-disease-may-have-a-more-negative-impact-than-the-covid-19-vaccine-especially-among-men?room_id=871-covid-19; Watson, R. E., T. B. Nelson, and A. L. Hsu, *Fertility considerations: The COVID-19 disease may have a more negative impact than the COVID-19 vaccine, especially among men*, Fertility and Sterility, March 19, 2021, https://www.fertstertdialog.com/posts/fertility-considerations-the-covid-19-disease-may-have-a-more-negative-impact-than-the-covid-19-vaccine-especially-among-men?room_id=871-covid-19; Chen, F., et al., *Effects of COVID-19 and mRNA vaccines on human fertility*, Human Reproduction, 2022. 37(1): 5–13; Bowman, C. J., et al., *Lack of effects on female fertility and prenatal and postnatal offspring development in rats with BNT162b2, a mRNA-based COVID-19 vaccine*, Reproductive Toxicology, 2021. 103:28–35.

31. Junkins, E., K. Lee, and K. Clancy, *Enhancing Knowledge through Engagement: Participation in an Unpaid Survey-Based Health Research*, International Congress of Qualitative Inquiry, virtual online meeting, May 2022; Lee, K., E. Junkins, and K. Clancy, *Menstrual experiences after COVID-19 vaccination in a non-menstruating gender diverse sample* (poster presentation), Experimental Biology, 2022. 36(S1), https://doi.org/10.1096/fasebj.2022.36.S1.R5912; Lee, K., et al., *Measuring menstruation: Methodological difficulties in studying things we don't talk about* (poster presented at the Human Biology Association), American Journal of Human Biology, 2022. 34(S2): e23740; Lee, K. M. N., et al. *Investigating trends in those who experience menstrual bleeding changes after SARS-CoV-2 vaccination.* medRxiv, 2022: 2021.10.11.21264863; Lee, K. M. N., et al., *Investigating trends in those who experience menstrual bleeding changes after SARS-CoV-2 vaccination*, 2022: Science Advances, 8(29) DOI: 10.1126/sciadv.abm7201.

32. Van der Molen, R., et al., *Menstrual blood closely resembles the uterine immune microenvironment and is clearly distinct from peripheral blood*, Human Reproduction, 2014. 29(2): 303–14; Crona Guterstam, Y., et al., *The cytokine profile of menstrual blood*, Acta Obstetricia et Gynecologica Scandinavica, 2021. 100(2): 339–46.

33. Antoniak, S., *The coagulation system in host defense*, Research and Practice in Thrombosis and Haemostasis, 2018. 2(3): 549–57.

34. Metsios, G. S., R. H. Moe, and G. D. Kitas, *Exercise and inflammation*, Best Practice and Research Clinical Rheumatology, 2020. 34(2): 101504.

Chapter 5: Big Little Stressors

1. Gordy, C., *Anita Hill Defends Her Legacy*, The Root, October 18, 2011, https://www.theroot.com/anita-hill-defends-her-legacy-1790866401.

2. Segers, G., *Here Are Some of the Questions Anita Hill Answered in 1991*, CBS News, September 19, 2018, https://www.cbsnews.com/news/here-are-some-of-the-questions-anita-hill-fielded-in-1991/.

3. Cortina, L., and V. Magley, *Raising voice, risking retaliation: Events following interpersonal mistreatment in the workplace*, Journal of Occupational Health Psychology, 2003. 8(4): 247.

4. Rocha Beardall, T., *Settler simultaneity and anti-Indigenous racism at land-grant universities*, Sociology of Race and Ethnicity, 2022. 8(1): 197–212; Embrick, D., and J. Williams, *Civility for*

Whom? Inside Higher Education, November 16, 2018, https://www.insidehighered.com/advice/2018/11/16/when-calls-civility-are-attempts-silence-messenger-opinion; Sugrue, T., *White America's Age-Old, Misguided Obsession with Civility,* New York Times, June 29, 2018, A23; Mills, C. W., *The Racial Contract.* 1997: Cornell University Press; Cortina, L. M., and V. J. Magley, *Patterns and profiles of response to incivility in the workplace,* Journal of Occupational Health Psychology, 2009. 14(3): 272.

5. Norman, A., *Ask Me about My Uterus: A Quest to Make Doctors Believe in Women's Pain.* 1st trade paperback ed. 2018: Bold Type Books.

6. Amico, J. R., et al., *"I wish they could hold on a little longer": Physicians' experiences with requests for early IUD removal,* Contraception, 2017. 96(2): 106–10.

7. Roberts, T. A., and S. Hansen, *Association of hormonal contraception with depression in the postpartum period,* Contraception, 2017. 96(6): 446–52; Behre, H. M., et al., *Efficacy and safety of an injectable combination hormonal contraceptive for men,* Journal of Clinical Endocrinology and Metabolism, 2016. 101(12): 4779–88; Oreskes, N. *Science awry.* In *Why Trust Science?* ed. N. Oreskes, pp. 69–146. 2019: Princeton University Press.

8. Alson, J. G., et al., *Incorporating measures of structural racism into population studies of reproductive health in the United States: A narrative review,* Health Equity, 2021. 5(1): 49–58; Dressler, W. W., K. S. Oths, and C. C. Gravlee. *Race and ethnicity in public health research: models to explain health disparities.* Annual Review of Anthropology, 2005. 34:231–52.

9. Ellison, P., et al., *Moderate anxiety, whether acute or chronic, is not associated with ovarian suppression in healthy, well-nourished, Western women,* American Journal of Physical Anthropology, 2007. 134(4): 513–19.

10. Bracha, H. S., *Freeze, flight, fight, fright, faint: Adaptationist perspectives on the acute stress response spectrum,* CNS Spectrums, 2004. 9(9): 679–85; Maack, D. J., E. Buchanan, and J. Young, *Development and psychometric investigation of an inventory to assess fight, flight, and freeze tendencies: The fight, flight, freeze questionnaire,* Cognitive Behaviour Therapy, 2015. 44(2): 117–27; Webster, V., P. Brough, and K. Daly, *Fight, flight or freeze: Common responses for follower coping with toxic leadership,* Stress and Health, 2016. 32(4): 346–54.

11. Wrangham, R. W., *Evolution of coalitionary killing,* Yearbook of Physical Anthropology, 1999. 42:1–30; Muller, M. N., and R. W. Wrangham, *Sexual Coercion in Primates and Humans.* 2009: Harvard University Press; Arseneau-Robar, T.J.M., et al., *Male monkeys use punishment and coercion to de-escalate costly intergroup fights,* Proceedings of the Royal Society B: Biological Sciences, 2018. 285(1880): 20172323.

12. Hrdy, S. B., *The Woman That Never Evolved.* 1981: Harvard University Press; Hrdy, S. B., *Mother Nature: A History of Mothers, Infants, and Natural Selection.* 1999: Pantheon.

13. Taylor, S. E., *Tend and befriend: Biobehavioral bases of affiliation under stress,* Current Directions in Psychological Science, 2006. 15(6): 273–77; Rodrigues, M. A. *Stress and sociality in a patrilocal primate: Do female spider monkeys tend-and-befriend?* PhD diss., Ohio State University, 2013.

14. Cortina, L. M., M. S. Hershcovis, and K. B. Clancy, *The embodiment of insult: A theory of biobehavioral response to workplace incivility,* Journal of Management, 2021. 48(3): 738–63.

15. Duboscq, J., et al., *The function of postconflict interactions: New prospects from the study of a tolerant species of primate,* Animal Behaviour, 2014. 87:107–20; Katsu, N., K. Yamada, and

M. Nakamichi, *Functions of post-conflict affiliation with a bystander differ between aggressors and victims in Japanese macaques*, Ethology, 2018. 124(2): 94–104; Taylor, S. E. *Tend and befriend: Biobehavioral bases of affiliation under stress.* Current Directions in Psychological Science, 2006. **15**(6): 273–77.

16. Rodrigues, M. A., R. Mendenhall, and K. Clancy, *"There's realizing, and then there's realizing": How social support can counter gaslighting of women of color scientists*, Journal of Women and Minorities in Science and Engineering, 2021. 27(2): 1–23.

17. Major, B., W. J. Quinton, and S. K. McCoy, *Antecedents and consequences of attributions to discrimination: Theoretical and empirical advances.* 2002. 34:251–330; Rodrigues, M. A., R. Mendenhall, and K. Clancy. *"There's realizing, and then there's realizing": How social support can counter gaslighting of women of color scientists.* Journal of Women and Minorities in Science and Engineering, 2021. **27**(2): 1–23.

18. Matheson, K., and H. Anisman, *Anger and shame elicited by discrimination: Moderating role of coping on action endorsements and salivary cortisol*, European Journal of Social Psychology, 2009. 39(2): 163–85.

19. Palm-Fischbacher, S., and U. Ehlert, *Dispositional resilience as a moderator of the relationship between chronic stress and irregular menstrual cycle*, Journal of Psychosomatic Obstetrics and Gynecology, 2014. 35(2): 42–50; Jacobs, M. B., R. D. Boynton-Jarrett, and E. W. Harville, *Adverse childhood event experiences, fertility difficulties and menstrual cycle characteristics*, Journal of Psychosomatic Obstetrics and Gynecology, 2015. 36(2): 46–57; Allsworth, J., et al., *The influence of stress on the menstrual cycle among newly incarcerated women*, Women's Health Issues, 2007. 17(4): 202–9; Natt, A. M., F. Khalid, and S. S. Sial, *Relationship between examination stress and menstrual irregularities among medical students of Rawalpindi Medical University*, Journal of Rawalpindi Medical College, 2018. 22(S–1): 44–47.

20. Dinh, T., et al., *Endocrinological effects of social exclusion and inclusion: Experimental evidence for adaptive regulation of female fecundity*, Hormones and Behavior, 2021. 130:104934.

21. Vitzthum, V. J., et al., *A prospective study of early pregnancy loss in humans*, Fertility and Sterility, 2006. 86(2): 373–79.

22. Banis, S., and M. M. Lorist, *The combined effects of menstrual cycle phase and acute stress on reward-related processing*, Biological Psychology, 2017. 125:130–45.

23. Duchesne, A., and J. C. Pruessner, *Association between subjective and cortisol stress response depends on the menstrual cycle phase*, Psychoneuroendocrinology, 2013. 38(12): 3155–59.

24. McKittrick, K., *The smallest cell remembers a sound*, in *Dear Science and Other Stories*. 2021: Duke University Press.

25. Ryu, A., and T.-H. Kim, *Premenstrual syndrome: A mini review*, Maturitas, 2015. 82(4): 436–40; Reid, R. L., and C. N. Soares, *Premenstrual dysphoric disorder: Contemporary diagnosis and management*, Journal of Obstetrics and Gynaecology Canada, 2018. 40(2): 215–23; Payne, J. L., and J. Maguire, *Pathophysiological mechanisms implicated in postpartum depression*, Frontiers in Neuroendocrinology, 2019. 52:165–80.

26. Halbreich, U., and S. Karkun, *Cross-cultural and social diversity of prevalence of postpartum depression and depressive symptoms*, Journal of Affective Disorders, 2006. 91(2–3): 97–111; Halbreich, U., et al., *The prevalence, impairment, impact, and burden of premenstrual dysphoric disorder (PMS/PMDD)*, Psychoneuroendocrinology, 2003. 28:1–23; Halbreich, U., et al., *The*

prevalence, impairment, impact, and burden of premenstrual dysphoric disorder (PMS/PMDD), Psychoneuroendocrinology, 2003. 28:1–23.

27. Matsumoto, T., et al., *Comparison between retrospective premenstrual symptoms and prospective late-luteal symptoms among college students,* Gynecological and Reproductive Endocrinology and Metabolism, 2021. 2(1): 31–41; Eisenlohr-Moul, T. A., et al., *Toward the reliable diagnosis of DSM-5 premenstrual dysphoric disorder: The Carolina Premenstrual Assessment Scoring System (C-PASS),* American Journal of Psychiatry, 2017. 174(1): 51–59.

28. Ussher, J. M., *The role of premenstrual dysphoric disorder in the subjectification of women,* Journal of Medical Humanities, 2003. 24(1): 131–46.

29. Ussher, J. M., and J. Perz, *Empathy, egalitarianism and emotion work in the relational negotiation of PMS: The experience of women in lesbian relationships,* Feminism and Psychology, 2008. 18(1): 87–111.

30. Pilver, C. E., et al., *Exposure to American culture is associated with premenstrual dysphoric disorder among ethnic minority women,* Journal of Affective Disorders, 2011. 130(1–2): 334–41.

31. Pilver, C. E., et al., *Lifetime discrimination associated with greater likelihood of premenstrual dysphoric disorder,* Journal of Women's Health, 2011. 20(6): 923–31.

32. Eisenlohr-Moul, T. A., et al., *Are there temporal subtypes of premenstrual dysphoric disorder? Using group-based trajectory modeling to identify individual differences in symptom change,* Psychological Medicine, 2020. 50(6): 964–72.

33. Martinez, P. E., et al., *5 α-reductase inhibition prevents the luteal phase increase in plasma allopregnanolone levels and mitigates symptoms in women with premenstrual dysphoric disorder,* Neuropsychopharmacology, 2016. 41(4): 1093–102; Dubey, N., et al., *The ESC/E (Z) complex, an effector of response to ovarian steroids, manifests an intrinsic difference in cells from women with premenstrual dysphoric disorder,* Molecular Psychiatry, 2017. 22(8): 1172–84.

34. Eisenlohr-Moul, T. A., et al., *Histories of abuse predict stronger within-person covariation of ovarian steroids and mood symptoms in women with menstrually related mood disorder,* Psychoneuroendocrinology, 2016. 67:142–52.

35. Loman, M. M., and M. R. Gunnar, *Early experience and the development of stress reactivity and regulation in children,* Neuroscience and Biobehavioral Reviews, 2010. 34(6): 867–76; Gunnar, M., and K. Quevedo, *The neurobiology of stress and development,* Annual Review of Psychology, 2007. 58(1): 145–73.

36. Timothy, A., et al., *Influence of early adversity on cortisol reactivity, SLC6A4 methylation and externalizing behavior in children of alcoholics,* Progress in Neuro-Psychopharmacology and Biological Psychiatry, 2019. 94:109649.

37. Goldman-Mellor, S., M. Hamer, and A. Steptoe, *Early-life stress and recurrent psychological distress over the lifecourse predict divergent cortisol reactivity patterns in adulthood,* Psychoneuroendocrinology, 2012. 37(11): 1755–68.

38. Ellis, B. J., et al., *Quality of early family relationships and the timing and tempo of puberty: Effects depend on biological sensitivity to context,* Development and Psychopathology, 2011. 23(1): 85.

39. Ellis, B. J., and M. J. Essex, *Family environments, adrenarche, and sexual maturation: A longitudinal test of a life history model,* Child Development, 2007. 78(6): 1799–817.

40. Felitti, V. J., et al., *Relationship of childhood abuse and household dysfunction to many of the leading causes of death in adults: The Adverse Childhood Experiences (ACE) study,* American

Journal of Preventive Medicine, 1998. 14(4): 245–58; Dube, S. R., et al., *The impact of adverse childhood experiences on health problems: Evidence from four birth cohorts dating back to 1900*, Preventive Medicine, 2003. 37(3): 268–77.

41. Edwards, V. J., et al., *Relationship between multiple forms of childhood maltreatment and adult mental health in community respondents: Results from the adverse childhood experiences study*, American Journal of Psychiatry, 2003. 160(8): 1453–60; Barnett, M. L., et al., *Implications of adverse childhood experiences screening on behavioral health services: A scoping review and systems modeling analysis*, American Psychologist, 2021. 76(2): 364; Steele, H., et al., *Adverse childhood experiences, poverty, and parenting stress*, Canadian Journal of Behavioural Science/Revue Cana; dienne des sciences du comportement, 2016. 48(1): 32; Pretty, C., et al., *Adverse childhood experiences and the cardiovascular health of children: A cross-sectional study*, BMC Pediatrics, 2013. 13(1): 1–8; Felitti, V. J., et al. *Relationship of childhood abuse and household dysfunction to many of the leading causes of death in adults*245–58; Dube, S. R., et al. *The impact of adverse childhood experiences on health problems* 268–77.

42. Negriff, S., *ACEs are not equal: Examining the relative impact of household dysfunction versus childhood maltreatment on mental health in adolescence*, Social Science and Medicine, 2020. 245:112696.

43. Hosseini-Kamkar, N., C. Lowe, and J. B. Morton, *The differential calibration of the HPA as a function of trauma versus adversity: A systematic review and p-curve meta-analyses*, Neuroscience and Biobehavioral Reviews, 2021. 127:154–330.

44. Rogers-LaVanne, M. P., *Variation in age at menarche and adult reproductive function: The role of energetic and psychosocial stressors*. PhD diss., University of Illinois Urbana-Champaign, 2018.

45. Rogers, M., *Variation in age at menarche and reproductive function in the United States and Poland*. PhD diss., University of Illinois, 2018.

46. Jacobs, M. B., R. D. Boynton-Jarrett, and E. W. Harville, *Adverse childhood event experiences, fertility difficulties and menstrual cycle characteristics*, Journal of Psychosomatic Obstetrics and Gynecology, 2015. 36(2): 46–57.

47. Latthe, P., et al., *WHO systematic review of prevalence of chronic pelvic pain: A neglected reproductive health morbidity*, BMC Public Health, 2006. 6(1): 177.

48. Meltzer-Brody, S., et al., *Trauma and posttraumatic stress disorder in women with chronic pelvic pain*, Obstetrics and Gynecology, 2007. 109(4): 902–8; Collett, B. J., et al., *A comparative study of women with chronic pelvic pain, chronic nonpelvic pain and those with no history of pain attending general practitioners*, Journal of Obstetrics and Gynaecology, 1998. 105(1): 87–92.

49. Park, S., et al., *Impact of childhood exposure to psychological trauma on the risk of psychiatric disorders and somatic discomfort: Single vs. multiple types of psychological trauma*, Psychiatry Research, 2014. 219(3): 443–49.

50. Vercellini, P., et al., *Association between endometriosis stage, lesion type, patient characteristics and severity of pelvic pain symptoms: A multivariate analysis of over 1000 patients*, Human Reproduction, 2007. 22(1): 266–71; Fauconnier, A., and C. Chapron, *Endometriosis and pelvic pain: Epidemiological evidence of the relationship and implications*, Human Reproduction Update, 2005. 11(6): 595–606.

51. Harris, H. R., et al., *Early life abuse and risk of endometriosis*, Human Reproduction, 2018. 33(9): 1657–68.

52. White, S., et al., *All the ACEs: A chaotic concept for family policy and decision-making?* Social Policy and Society, 2019. 18(3): 457–66.

53. Mendenhall, R., *The medicalization of poverty in the lives of low-income Black mothers and children,* Journal of Law, Medicine and Ethics, 2018. 46(3): 644–50.

54. Felitti, V. J., *How Childhood Trauma Can Make You a Sick Adult,* Bigthink, video, 2019, https://bigthink.com/u/vincent-felitti.

55. Lester, C., *How Childhood Trauma Affects Health* (blog), July 13, 2017, http://blogs.wgbh.org/innovation-hub/2017/7/13/felitti-ace/.

56. Daly, M., A. R. Sutin, and E. Robinson, *Perceived weight discrimination mediates the prospective association between obesity and physiological dysregulation: Evidence from a population-based cohort,* Psychological Science, 2019. 30(7): 1030–39; Tomiyama, A. J., et al. *How and why weight stigma drives the obesity "epidemic" and harms health.* BMC Medicine, 2018. 16(1): 123.

57. Sanders, K. M., et al., *Heightened cortisol response to exercise challenge in women with functional hypothalamic amenorrhea,* American Journal of Obstetrics and Gynecology, 2018. 218(2): 230. e1–e6.

58. Berga, S., et al., *Neuroendocrine aberrations in women with functional hypothalamic amenorrhea,* Journal of Clinical Endocrinology and Metabolism, 1989. 68(2): 301–8; Marcus, M. D., T. L. Loucks, and S. L. Berga, *Psychological correlates of functional hypothalamic amenorrhea,* Fertility and Sterility, 2001. 76(2): 310–16; Suh, B., et al., *Hypercortisolism in patients with functional hypothalamic-amenorrhea,* Journal of Clinical Endocrinology and Metabolism, 1988. 66(4): 733–39; Pauli, S. A., and S. L. Berga, *Athletic amenorrhea: Energy deficit or psychogenic challenge?* Annals of the New York Academy of Sciences, 2010. 1205(1): 33–38.

59. Strings, S. *Fearing the Black Body: The Racial Origins of Fat Phobia.* 2019: New York University Press.

60. Freizinger, M., et al., *The prevalence of eating disorders in infertile women,* Fertility and Sterility, 2010. 93(1): 72–78; Linna, M. S., et al., *Reproductive health outcomes in eating disorders,* International Journal of Eating Disorders, 2013. 46(8): 826–33; Audier-Bourgain, M., et al., *Eating disorders and sexuality: A quantitative study in a French medically assisted procreation course,* Brain and Behavior, e02196.

61. Czamanski-Cohen, J., et al., *Practice makes perfect: The effect of cognitive behavioral interventions during IVF treatments on women's perceived stress, plasma cortisol and pregnancy rates,* Journal of Reproductive Health and Medicine, 2016. 2:S21–S26.

62. Butler, L. D., et al., *Psychosocial predictors of resilience after the September 11, 2001 terrorist attacks,* Journal of Nervous and Mental Disease, 2009. 197(4): 266–73.

Chapter 6: The Future of Periods

1. Sanabria, E., *Introduction,* in *Plastic Bodies: Sex Hormones and Menstrual Suppression in Brazil.* 2016: Duke University Press.

2. Hasson, K. A., *Not a "real" period? Social and material constructions of menstruation,* Gender and Society, 2016. 30(6): 958–83; Sanabria, E. *Introduction.* In *Plastic Bodies*

3. Vostral, S. L. *Under Wraps: A History of Menstrual Hygiene Technology.* 2008: Lexington Books.

4. Leidy Sievert, L., *Should women menstruate? An evolutionary perspective on menstrual-suppressing oral contraceptives,* in *Evolutionary Medicine and Health: New Perspectives,* ed. W. Trevathan, E. O. Smith, and J. J. McKenna. 2008: Oxford University Press, pp. 181–95.

5. Bobel, C., *"Dignity can't wait": Building a bridge to human rights,* in *The Managed Body.* 2019: Springer, pp. 211–42; Luna, Z., and K. Luker, *Reproductive justice,* Annual Review of Law and Social Science, 2013. 9:327–52; Piepzna-Samarasinha, L. L., *Making space accessible is an act of love for our communities,* in *Care Work: Dreaming Disability Justice.* 2018: Arsenal Pulp Press.

6. O'Brien, M. H., *Being a scientist means taking sides,* Professional Biologist, 1993. 43(10): 706–8.

7. O'Brien, M. H. *Being a scientist means taking sides.* 706–8.

8. Liboiron, M. *Pollution Is Colonialism.* 2021: Duke University Press.

9. Piepzna-Samarasinha, L. L., *Crip emotional intelligence,* in Piepzna-Samarasinha, *Care Work.*

10. Mack, N., et al., *Strategies to improve adherence and continuation of shorter-term hormonal methods of contraception,* Cochrane Database of Systematic Reviews, 2019. 4(4): CD004317.

11. Triebwasser, J. E., et al., *Pharmacy claims data versus patient self-report to measure contraceptive method continuation,* Contraception, 2015. 92(1): 26–30.

12. Halpern, V., et al., *Strategies to improve adherence and acceptability of hormonal methods of contraception,* Cochrane Database of Systematic Reviews, 2013. 26(10): CD004317.

13. Geronimus, A. T., *Damned if you do: Culture, identity, privilege, and teenage childbearing in the United States,* Social Science and Medicine, 2003. 57(5): 881–93; Downing, R. A., T. A. LaVeist, and H. E. Bullock, *Intersections of ethnicity and social class in provider advice regarding reproductive health,* American Journal of Public Health, 2007. 97(10): 1803–7; Gomez, A. M., L. Fuentes, and A. Allina, *Women or LARC first? Reproductive autonomy and the promotion of long-acting reversible contraceptive methods,* Perspectives on Sexual and Reproductive Health, 2014. 46(3): 171–75; Dehlendorf, C., et al., *Recommendations for intrauterine contraception: A randomized trial of the effects of patients' race/ethnicity and socioeconomic status,* American Journal of Obstetrics and Gynecology, 2010. 203(4): 319. e1–e8.

14. Daud, S., and A. A. Ewies, *Levonorgestrel-releasing intrauterine system: Why do some women dislike it?,* Gynecological Endocrinology, 2008. 24(12): 686–90.

15. Grzanka, P. R., and E. Schuch, *Reproductive anxiety and conditional agency at the intersections of privilege: A focus group study of emerging adults' perception of long-acting reversible contraception,* Journal of Social Issues, 2020. 76(2): 270–313.

16. Bertotti, A. M., E. S. Mann, and S. A. Miner, *Efficacy as safety: Dominant cultural assumptions and the assessment of contraceptive risk,* Social Science and Medicine, 2021. 270:113547.

17. Strasser, J., et al., *Access to removal of long-acting reversible contraceptive methods is an essential component of high-quality contraceptive care,* Women's Health Issues, 2017. 27(3): 253–55.

18. Byrne, M., Twitter post, March 2, 2020, https://twitter.com/monicabyrne13/status/1234521738240765953?s=20.

19. Swehla, T., Twitter post, March 2, 2020, https://twitter.com/SwehlaTessa/status/1234532953675509760?s=20; Hamburg, S., Twitter post, March 2, 2020, https://twitter.com/sarahrhamburg/status/1235227052778938368?s=20; Hopkinson, N., Twitter post, March 2, 2020, https://twitter.com/Nalo_Hopkinson/status/1234526087348015104?s=20.

20. Perez-Lopez, F. R., et al., *Uterine or paracervical lidocaine application for pain control during intrauterine contraceptive device insertion: A meta-analysis of randomised controlled trials*, European Journal of Contraception and Reproductive Health Care, 2018. 23(3): 207–17.

21. Dahvana Headley, M., Twitter post, March 2, 2020, https://twitter.com/MARIADAHVANA/status/1234588409357721600?s=20.

22. Jones, C. E., *The pain of endo existence: Toward a feminist disability studies reading of endometriosis*, Hypatia, 2016. 31(3): 554–71.

23. Vercellini, P., et al., *Oral contraceptives and risk of endometriosis: A systematic review and meta-analysis*, Human Reproduction Update, 2010. 17(2): 159–70.

24. Tu, F. F., et al., *The influence of prior oral contraceptive use on risk of endometriosis is conditional on parity*, Fertility and sterility, 2014. 101(6): 1697–704.

25. Farland, L. V., et al., *Epidemiological and clinical risk factors for endometriosis*, in *Biomarkers for Endometriosis: State of the Art*, ed. T. D'Hooghe. 2019: Springer.

26. Di Donato, N., et al., *Prevalence of adenomyosis in women undergoing surgery for endometriosis*, European Journal of Obstetrics and Gynecology and Reproductive Biology, 2014. 181:289–93.

27. Evaristo, B., *Girl, Woman, Other*. 2019: Hamish Hamilton, an imprint of Penguin Random House UK.

28. Carlson, K. J., D. H. Nichols, and I. Schiff, *Indications for hysterectomy*, New England Journal of Medicine, 1993. 328(12): 856–60.

29. Rizk, B., et al., *Recurrence of endometriosis after hysterectomy*, Facts, Views and Vision in ObGyn, 2014. 6(4): 219.

30. Stewart, E. A., L. T. Shuster, and W. A. Rocca, *Reassessing hysterectomy*, Minnesota Medicine, 2012. 95(3): 36.

31. Altman, D., et al., *Hysterectomy and risk of stress-urinary-incontinence surgery: Nationwide cohort study*, Lancet, 2007. 370(9597): 1494–99.

32. Robinson, W. R., et al., *For U.S. Black women, shift of hysterectomy to outpatient settings may have lagged behind white women: A claims-based analysis, 2011–2013*, BMC Health Services Research, 2018. 17(526): 1–9.

33. Ahmad, S., and M. Leinung, *The response of the menstrual cycle to initiation of hormonal therapy in transgender men*, Transgender Health, 2017. 2(1): 176–79.

34. Toze, M., *The risky womb and the unthinkability of the pregnant man: Addressing trans masculine hysterectomy*, Feminism and Psychology, 2018. 28(2): 194–211; Rachlin, K., G. Hansbury, and S. T. Pardo, *Hysterectomy and oophorectomy experiences of female-to-male transgender individuals*, International Journal of Transgenderism, 2010. 12(3): 155–66.

35. Bretschneider, C. E., et al., *Complication rates and outcomes after hysterectomy in transgender men*, Obstetrics and Gynecology, 2018. 132(5): 1265–73.

36. Stubblefield, A., *"Beyond the pale": Tainted whiteness, cognitive disability, and eugenic sterilization*, Hypatia, 2007. 22(2): 162–81.

37. Freidenfelds, L., *The Modern Period: Menstruation in Twentieth-Century America*. 2009: Johns Hopkins University Press.

38. Aubeeluck, A., and M. Maguire, *The Menstrual Joy Questionnaire items alone can positively prime reporting of menstrual attitudes and symptoms*, Psychology of Women Quarterly, 2002. 26(2): 160–62.

39. Bobel, C., and B. Fahs, *From bloodless respectability to radical menstrual embodiment: Shifting menstrual politics from private to public*, Signs: Journal of Women in Culture and Society, 2020. 45(4): 955–83.

40. Gillies, V., et al., *Women's collective constructions of embodied practices through memory work: Cartesian dualism in memories of sweating and pain*, British Journal of Social Psychology, 2004. 43(1): 99–112.

41. Przybylo, E., and B. Fahs, *Feels and flows: On the realness of menstrual pain and cripping menstrual chronicity*, Feminist Formations, 2018. 30(1): 206–29.

42. McKinley, R., *The Blue Sword*. 1982: Greenwillow Books.

43. King, S., *The Stand*. 1978: Doubleday.

44. Yaszek, L., *The women history doesn't see: Recovering midcentury women's SF as a literature of social critique*, Extrapolation, 2004. 45(1): 34–51.

45. Shelley, M. W., *Frankenstein*. 1968, New York: Golden Press.

46. King, S., *Carrie*. 1974: Doubleday.

47. Willis, C., *Daisy in the sun*, in *Fire Watch: Memorable Tales by a Young Master*, ed. C. Willis. 1985, Bluejay Books.

48. Willis, C., *Even the queen*, in *Impossible Things*, ed. C. Willis. 1994: Bantam Spectra.

49. Zhu, Z., et al., *Mortality and morbidity of infants born extremely preterm at tertiary medical centers in China from 2010 to 2019*, JAMA Network Open, 2021. 4(5): e219382–e219382.

50. Banerjee, A., *Race and a transnational reproductive caste system: Indian transnational surrogacy*, Hypatia, 2014. 29(1): 113–28.

51. Schurr, C., *From biopolitics to bioeconomies: The ART of (re-) producing white futures in Mexico's surrogacy market*, Environment and Planning D: Society and Space, 2017. 35(2): 241–62.

52. Corey, J. S., *Leviathan Wakes*. 2011: Orbit.

53. Thompson, T., ed. *The Murders of Molly Southbourne*. 2017: Tor; Thompson, T., ed. *The Survival of Molly Southbourne*. 2019: Tom Doherty.

54. Dhaliwal, A., *Woman World*. 2018: Drawn and Quarterly.

55. Spruck Wrigley, S., *Space Travel Loses Its Allure When You've Lost Your Moon Cup*, Flash Fiction Online, October 2015, https://www.flashfictiononline.com/article/space-travel-loses-its-allure-when-youve-lost-your-moon-cup/.

56. Fitzwater, A., *Logistics*, Clarkesworld, no. 139, April 2018, https://clarkesworldmagazine.com/audio_04_18c/.

57. Schalk, S., *The future of bodyminds, bodyminds of the future*, in *Bodyminds Reimagined: (Dis)ability, Race, and Gender in Black Women's Speculative Fiction*. 2018: Duke University Press.

58. Schalk, S., *Introduction*, in Schalk, *Bodyminds Reimagined*.

59. Butler, O. E., *Parable of the Sower*. Earthseed. 1993: Aspect Books.

Epilogue

1. Liboiron, M. *Pollution Is Colonialism*. 2021: Duke University Press.

2. Al-Ansari, A. M., et al., *Bioaccumulation of the pharmaceutical 17a-ethinylestradiol in shorthead redhorse suckers* (Moxostoma macrolepidotum) *from the St. Clair River, Canada,*

Environmental Pollution, 2010. 158(8): 2566–71; Adeel, M., et al., *Environmental impact of estrogens on human, animal and plant life: A critical review*, Environment International, 2017. 99:107–19.

3. Cortina, L. M., *Unseen injustice: Incivility as modern discrimination in organizations*, Academy of Management Review, 2008. 33(1): 55–75; Cortina, L. M., et al., *Selective incivility as modern discrimination in organizations evidence and impact*, Journal of Management, 2013. 39(6): 1579–605; Cortina, L. M., et al., *Researching rudeness: The past, present, and future of the science of incivility*, Journal of Occupational Health Psychology, 2017. 22(3): 299; Barthelemy, R., M. McCormick, and C. Henderson, *Understanding Women's Gendered Experiences in Physics and Astronomy through Microaggressions*, in *2014 Physics Education Research Conference Proceedings*, ed. P. Engelhardt, A. Churukian, and D. Jones. 2015: American Association of Physics Teachers, pp. 35–38; Basford, T. E., L. R. Offermann, and T. S. Behrend, *Do you see what I see? Perceptions of gender microaggressions in the workplace*, Psychology of Women Quarterly, 2014. 38(3): 340–49; Fatima, S., *On the edge of knowing: Microaggression and epistemic uncertainty as a woman of color*, in *Surviving Sexism in Academia*, ed. K. Cole and H. Hassel. 2017: Routledge; Lewis, J. A., et al., *Racial microaggressions and sense of belonging at a historically white university*, American Behavioral Scientist, 2019. 65(8): 1049–71; Nadal, K. L., et al., *Sexual orientation and transgender microaggressions*, in Microaggressions and Marginality: Manifestation, Dynamics, and Impact, ed. Derald Wing Sue. 2010: Wiley, pp. 217–40.

4. Thompson, C., et al., *10 Dead in Buffalo Supermarket Attack Police Call Hate Crime*, AP News, May 14, 2022.

5. Johnson, M. E., *Menstrual justice*, UC Davis Law Review, 2019. 53:1.

6. White Junod, S., and L. Marks, *Women's trials: The approval of the first oral contraceptive pill in the United States and Great Britain*, Journal of the History of Medicine, 2002. 57:117–60.

7. Skloot, R., *The Immortal Life of Henrietta Lacks*. 2011: Crown.

8. Layne, L. L., *Introduction*, in *Feminist Technology*, ed. L. L. Layne, S. Vostral, and K. Boyer. 2010: University of Illinois Press.

9. Liboiron, M., M. Tironi, and N. Calvillo, *Toxic politics: Acting in a permanently polluted world*, Social Studies of Science, 2018. 48(3): 331–49; Liboiron, M. *Pollution Is Colonialism*.

10. Wilson, K., B. Sonenstein, and M. Kaba, *Hope Is a Discipline*, interview by Kim Wilson and Brian Sonenstein, Beyond Prisons, January 2018, https://www.beyond-prisons.com/home/hope-is-a-discipline-feat-mariame-kaba.

REFERENCES

Abalos, E., et al. *Global and regional estimates of preeclampsia and eclampsia: A systematic review.* European Journal of Obstetrics and Gynecology and Reproductive Biology, 2013. **170**(1): 1–7.

Abimbola, S. *The foreign gaze: Authorship in academic global health.* BMJ Specialist Journals, 2019: e002068.Adeel, M., et al. *Environmental impact of estrogens on human, animal and plant life: A critical review.* Environment International, 2017. **99**:107–19.

Ahmad, S., and M. Leinung. *The response of the menstrual cycle to initiation of hormonal therapy in transgender men.* Transgender Health, 2017. **2**(1): 176–79.

Ahmed, S., and A. Avasarala. *Urinary tract infections (UTI) among adolescent girls in rural Karimnagar District, AP KAP STUDY.* Indian Journal of Preventive and Social Medicine, 2008. **39**(1): 6–9.

Al-Ansari, A. M., et al. *Bioaccumulation of the pharmaceutical 17a-ethinylestradiol in shorthead redhorse suckers* (Moxostoma macrolepidotum) *from the St. Clair River, Canada.* Environmental Pollution, 2010. **158**(8): 2566–71.

Albamonte, M. I., et al. *The infant and pubertal human ovary: Balbiani's body-associated VASA expression, immunohistochemical detection of apoptosis-related BCL2 and BAX proteins, and DNA fragmentation.* Human Reproduction, 2013. **28**(3): 698–706.

Allsworth, J., et al. *The influence of stress on the menstrual cycle among newly incarcerated women.* Women's Health Issues, 2007. **17**(4): 202–9.

Almeling, R. *GUYnecology: The Missing Science of Men's Reproductive Health.* 2020: University of California Press.

Almeling, R. *Sex, sperm, and fatherhood.* In *GUYnecology: The Missing Science of Men's Reproductive Health,* chap. 5. 2020: University of California Press.

Alson, J. G., et al. *Incorporating measures of structural racism into population studies of reproductive health in the United States: A narrative review.* Health Equity, 2021. **5**(1): 49–58.

Altman, D., et al. *Hysterectomy and risk of stress-urinary-incontinence surgery: Nationwide cohort study.* Lancet, 2007. **370**(9597): 1494–99.

Amatya, P., et al. *Practice and lived experience of menstrual exiles (Chhaupadi) among adolescent girls in far-western Nepal.* PloS One, 2018. **13**(12): e0208260.

Amico, J. R., et al. *Access to IUD removal: Data from a mystery-caller study.* Contraception, 2020. **101**(2): 122–29.

Amico, J. R., et al. *"I wish they could hold on a little longer": Physicians' experiences with requests for early IUD removal.* Contraception, 2017. **96**(2): 106–10.

Amico, J. R., et al. *"She just told me to leave it": Women's experiences discussing early elective IUD removal.* Contraception, 2016. **94**(4): 357–61.

Amico, J. R., et al. *"$231 . . . to pull a string!!!" American IUD users' reasons for IUD self-removal: An analysis of internet forums.* Contraception, 2020. **101**(6): 393–98.

Amos, V., and P. Parmar. *Challenging imperial feminism.* Feminist Review, 1984. **17**(1): 3–19.

Anderson, K. *Life Stages and Native Women: Memory, Teachings, and Story Medicine.* Critical Studies in Native History, ed. J. Brownlie. 2011: University of Manitoba Press.

Antoniak, S. *The coagulation system in host defense.* Research and Practice in Thrombosis and Haemostasis, 2018. **2**(3): 549–57.

Arseneau-Robar, T.J.M., et al. *Male monkeys use punishment and coercion to de-escalate costly intergroup fights.* Proceedings of the Royal Society B—Biological Sciences, 2018. **285**(1880): 20172323.

Aschwanden, C. *Good to Go: What the Athlete in All of Us Can Learn from the Strange Science of Recovery.* 2019: W. W. Norton.

Ashley-Montagu, M. *Physiology and the origins of the menstrual prohibitions.* Quarterly Review of Biology, 1940. **15**(2): 211–20.

Aubeeluck, A., and M. Maguire. *The Menstrual Joy questionnaire items alone can positively prime reporting of menstrual attitudes and symptoms.* Psychology of Women Quarterly, 2002. **26**(2): 160–62.

Audier-Bourgain, M., et al. *Eating disorders and sexuality: A quantitative study in a French medically assisted procreation course.* Brain and Behavior, e02196.

Aycock, L., et al. *Sexual harassment reported by undergraduate female physicists.* Physical Review Physics Education Research, 2019. **15**:010121.

Baerwald, A. R., G. P. Adams, and R. A. Pierson. *Characterization of ovarian follicular wave dynamics in women.* Biology of Reproduction, 2003. **69**(3): 1023–31.

Baerwald, A. R., et al. *Age-related changes in luteal dynamics: Preliminary associations with antral follicular dynamics and hormone production during the human menstrual cycle.* Menopause, 2018. **25**(4): 399–407.

Baerwald, A. R., O. A. Olatunbosun, and R. A. Pierson. *Effects of oral contraceptives administered at defined stages of ovarian follicular development.* Fertility and Sterility, 2006. **86**(1): 27–35.

Baerwald, A. R., O. A. Olatunbosun, and R. A. Pierson. *Ovarian follicular development is initiated during the hormone-free interval of oral contraceptive use.* Contraception, 2004. **70**(5): 371–77.

Bajaj, M., A. Ghaffar-Kucher, and K. Desai. *Brown bodies and xenophobic bullying in US schools: Critical analysis and strategies for action.* Harvard Educational Review, 2016. **86**(4): 481–505.

Baker, K., M. P. Eichhorn, and M. Griffiths. *Decolonizing field ecology.* Biotropica, 2019. **51**(3): 288–92.

Baker, L. D. *Anthropology and the Racial Politics of Culture.* 2010: Duke University Press.

Baldy, C. R. *We Are Dancing for You: Native Feminisms and the Revitalization of Women's Coming-of-Age Ceremonies.* Indigenous Confluences, ed. C. Cote and C. Thrush. 2018: University of Washington Press.

Banerjee, A. *Race and a transnational reproductive caste system: Indian transnational surrogacy.* Hypatia, 2014. **29**(1): 113–28.

Banis, S., and M. M. Lorist. *The combined effects of menstrual cycle phase and acute stress on reward-related processing.* Biological Psychology, 2017. **125**:130–45.

Barnett, M. L., et al. *Implications of adverse childhood experiences screening on behavioral health services: A scoping review and systems modeling analysis.* American Psychologist, 2021. **76**(2): 364.

Barres, B. *Does gender matter?* Nature, 2006. **442**:133–36.

Barthelemy, R., M. McCormick, and C. Henderson. *Understanding women's gendered experiences in physics and astronomy through microaggressions.* In *2014 Physics Education Research Conference Proceedings*, ed. P. Engelhardt, A. Churukian, and D. Jones, pp. 35–38. 2015: American Association of Physics Teachers.

Basford, T. E., L. R. Offermann, and T. S. Behrend. *Do you see what I see? Perceptions of gender microaggressions in the workplace.* Psychology of Women Quarterly, 2014. **38**(3): 340–49.

Beeler, J., F. Varricchio, and R. Wise. *Thrombocytopenia after immunization with measles vaccines: Review of the vaccine adverse events reporting system (1990 to 1994).* Pediatric Infectious Disease Journal, 1996. **15**(1): 88–90.

Behre, H. M., et al. *Efficacy and safety of an injectable combination hormonal contraceptive for men.* Journal of Clinical Endocrinology and Metabolism, 2016. **101**(12): 4779–88.

Beisel, N., and T. Kay. *Abortion, race, and gender in nineteenth-century America.* American Sociological Review, 2004. **69**(4): 498–518.

Bekker, S. *On the History of (the Segregation of) Women's Sport, Holding Space*, March 20, 2022.

Bentley, G. R., et al. *Women's strategies to alleviate nutritional stress in a rural African society.* Social Science and Medicine, 1999. **48**(2): 149–62.

Bentley, G. R., A. M. Harrigan, and P. T. Ellison. *Ovarian cycle length and days of menstruation of Lese horticulturalists (abstract).* American Journal of Physical Anthropology, 1990. **81**:193–94.

Berdahl, J. L., et al. *Work as a masculinity contest.* Journal of Social Issues, 2018. **74**(3): 422–48.

Berga, S., et al. *Neuroendocrine aberrations in women with functional hypothalamic amenorrhea.* Journal of Clinical Endocrinology and Metabolism, 1989. **68**(2): 301–8.

Bertotti, A. M., E. S. Mann, and S. A. Miner. *Efficacy as safety: Dominant cultural assumptions and the assessment of contraceptive risk.* Social Science and Medicine, 2021. **270**:113547.

Besse, M., N. M. Lampe, and E. S. Mann. *Focus: Sex and reproduction: Experiences with achieving pregnancy and giving birth among transgender men: A narrative literature review.* Yale Journal of Biology and Medicine, 2020. **93**(4): 517.

Bianco, P., P. Gehron Robey, and P. Simmons. *Mesenchymal stem cells: Revisiting history, concepts, and assays.* Cell Stem Cell, 2008. **2**(4): 313–19.

Bildhauer, B. *Medieval Blood.* 2006: University of Wales Press.

Black, P., and U. Sharma. *Men are real, women are "made up": Beauty therapy and the construction of femininity.* Sociological Review, 2001. **49**(1): 100–116.

Blackless, M., et al. *How sexually dimorphic are we? Review and synthesis.* American Journal of Human Biology, 2000. **12**(2): 151–66.

Bloomfield, S. F., et al. *Time to abandon the hygiene hypothesis: New perspectives on allergic disease, the human microbiome, infectious disease prevention and the role of targeted hygiene.* Perspectives in Public Health, 2016. **136**(4): 213–24.

Blum, C., et al. *Low-grade inflammation and estimates of insulin resistance during the menstrual cycle in lean and overweight women.* Journal of Clinical Endocrinology and Metabolism, 2005. **90**(6): 3230–35.

Bobel, C. *"Dignity can't wait": Building a bridge to human rights.* In *The Managed Body*, pp. 211–42. 2019: Springer.

Bobel, C. *The Managed Body: Developing Girls and Menstrual Health in the Global South.* 2019: Palgrave Macmillan.

Bobel, C., and B. Fahs. *From bloodless respectability to radical menstrual embodiment: Shifting menstrual politics from private to public.* Signs: Journal of Women in Culture and Society, 2020. **45**(4): 955–83.

Bonaparte, A. D. *Physicians' discourse for establishing authoritative knowledge in birthing work and reducing the presence of the granny midwife.* Journal of Historical Sociology, 2015. **28**(2): 166–94.

Bove, R. B., C. R. Valeggia, and P. T. Ellison. *Girl helpers and time allocation of nursing women among the Toba of Argentina.* Human Nature, 2002. **13**(4): 457–72.

Bowman, C. J., et al. *Lack of effects on female fertility and prenatal and postnatal offspring development in rats with BNT162b2, a mRNA-based COVID-19 vaccine.* Reproductive Toxicology, 2021. **103**:28–35.

Bracha, H. S. *Freeze, flight, fight, fright, faint: Adaptationist perspectives on the acute stress response spectrum.* CNS Spectrums, 2004. **9**(9): 679–85.

Brant, J. *Duel in the Sun. Runner's World*, April 4, 2004.

Bretschneider, C. E., et al. *Complication rates and outcomes after hysterectomy in transgender men.* Obstetrics and Gynecology, 2018. **132**(5): 1265–73.

Brewis, A. A. *Stigma and the perpetuation of obesity.* Social Science and Medicine, 2014. **118**:152–58.

Brodkin, K., S. Morgen, and J. Hutchinson. *Anthropology as white public space?* American Anthropologist, 2011. **113**(4): 545–56.

Broocks, A., et al. *Cyclic ovarian function in recreational athletes.* Journal of Applied Physiology, 1990. **68**(5): 2083–86.

Brooks-Gunn, J., and D. N. Ruble. *Men's and women's attitudes and beliefs about the menstrual cycle.* Sex Roles, 1986. **14**(5–6): 287–99.

Brosens, I., et al. *The impact of uterine immaturity on obstetrical syndromes during adolescence.* American Journal of Obstetrics and Gynecology, 2017. **217**(5): 546–55.

Brosens, J. J., et al. *A role for menstruation in preconditioning the uterus for successful pregnancy.* American Journal of Obstetrics and Gynecology, 2009. **200**(6): 615. e1–e6.

Bryson, A., A. Koyama, and A. Hassan. *Addressing long-acting reversible contraception access, bias, and coercion: Supporting adolescent and young adult reproductive autonomy.* Current Opinion in Pediatrics, 2021. **33**(4): 345–53.

Buckley, T. *Menstruation and the power of Yurok women.* In *Blood Magic: The Anthropology of Menstruation*, ed. T. Buckley and A. Gottlieb. 1988: University of California Press.

Buckley, T. *Menstruation and the power of Yurok women: Methods in cultural reconstruction.* American Ethnologist, 1982. **9**(1): 47–60.

Buckley, T., and A. Gottlieb, eds. *Blood Magic: The Anthropology of Menstruation.* 1988: University of California Press.

Buckley, T., and A. Gottlieb. *A critical appraisal of theories of menstrual symbolism.* In *Blood Magic: The Anthropology of Menstruation,* ed. T. Buckley and A. Gottlieb. 1988: University of California Press.

Burrows, A., and S. Johnson. *Girls' experiences of menarche and menstruation.* Journal of Reproductive and Infant Psychology, 2005. **23**(3): 235–49.

Butler, L. D., et al. *Psychosocial predictors of resilience after the September 11, 2001 terrorist attacks.* Journal of Nervous and Mental Disease, 2009. **197**(4): 266–73.

Butler, O. E. *Parable of the Sower.* Earthseed. 1993: Aspect Books.

Byrne, M. Twitter post, March 2, 2020. https://twitter.com/monicabyrne13/status/12345217 38240765953?s=20.

Cain, M. *I Was the Fastest Girl in America, until I Joined Nike.* New York Times, November 7, 2019.

Carlson, K. J., D. H. Nichols, and I. Schiff. *Indications for hysterectomy.* New England Journal of Medicine, 1993. **328**(12): 856–60.

Caron, C. *What Women Need to Know about the Covid Vaccine.* New York Times, September 13, 2021.

Chen, F., et al. *Effects of COVID-19 and mRNA vaccines on human fertility.* Human Reproduction, 2022. **37**(1): 5–13.

Chen, X., et al. *Prevalence of polycystic ovary syndrome in unselected women from southern China.* European Journal of Obstetrics and Gynecology and Reproductive Biology, 2008. **139**(1): 59–64.

Cheng, C. Y., K. Yang, and S. R. Liou. *Taiwanese adolescents' gender differences in knowledge and attitudes towards menstruation.* Nursing and Health Sciences, 2007. **9**(2): 127–34.

Cheyney, M. J. *Homebirth as systems-challenging praxis: Knowledge, power, and intimacy in the birthplace.* Qualitative Health Research, 2008. **18**(2): 254–67.

Clancy, K. B. H. Twitter post, 2018. https://twitter.com/KateClancy/status/948788429 860605952.

Clancy, K. B. H. *Unexpected luteal endometrial decline in a healthy rural Polish population.* European Journal of Obstetrics and Gynecology and Reproductive Biology, 2007. **134**(1): 133–34.

Clancy, K. B. H., A. Baerwald, and R. Pierson. *Systemic inflammation is associated with ovarian follicular dynamics during the human menstrual cycle.* PLoS One, 2013. **8**(5): e64807.

Clancy, K. B. H., L. M. Cortina, and A. R. Kirkland. *Opinion: Use science to stop sexual harassment in higher education.* Proceedings of the National Academy of Sciences, 2020. **117**(37): 22614–18.

Clancy, K. B. H., and J. Davis. *Soylent is people and WEIRD is white: Biological anthropology, whiteness, and the limits of the WEIRD.* Annual Review of Anthropology, 2019. **48**:169–86.

Clancy, K. B. H., et al. *Double jeopardy in astronomy and planetary science: Women of color face greater risks of gendered and racial harassment.* JGR Planets, 2017. **122**(7): 1610–23.

Clancy, K. B. H., et al. *Endometrial thickness is not independent of luteal phase day in a rural Polish population.* Anthropological Science, 2009. **117**(3): 157–63.

Clancy, K. B. H., et al. *Relationships between biomarkers of inflammation, ovarian steroids, and age at menarche in a rural Polish sample.* American Journal of Human Biology, 2013. **25**(3): 389–98.

Clancy, K. B. H., et al. *Survey of Academic Field Experiences (SAFE): Trainees report harassment and assault.* PLoS ONE, 2014. **9**(7): 1–9.

Clancy, K. B. H., I. Nenko, and G. Jasienska. *Menstruation does not cause anemia: Endometrial thickness correlates positively with erythrocyte count and hemoglobin concentration in premenopausal women.* American Journal of Human Biology, 2006. **18**(5): 710–13.

Clarke, E. H. *Sex in Education; or, A Fair Chance for the Girls.* 1873: James R. Osgood.

Collett, B. J., et al. *A comparative study of women with chronic pelvic pain, chronic nonpelvic pain and those with no history of pain attending general practitioners.* Journal of Obstetrics and Gynaecology, 1998. **105**(1): 87–92.

Corey, J. S. *Leviathan Wakes.* 2011: Orbit.

Cortina, L. M. *Unseen injustice: Incivility as modern discrimination in organizations.* Academy of Management Review, 2008. **33**(1): 55–75.

Cortina, L. M., et al. *Researching rudeness: The past, present, and future of the science of incivility.* Journal of Occupational Health Psychology, 2017. **22**(3): 299.

Cortina, L. M., et al. *Selective incivility as modern discrimination in organizations evidence and impact.* Journal of Management, 2013. **39**(6): 1579–605.

Cortina, L. M., M. S. Hershcovis, and K. B. Clancy. *The embodiment of insult: A theory of biobehavioral response to workplace incivility.* Journal of Management, 2021. **48**(3): 738–63.

Cortina, L. M., and V. J. Magley. *Patterns and profiles of response to incivility in the workplace.* Journal of Occupational Health Psychology, 2009. **14**(3): 272.

Cortina, L. M., and V. J. Magley. *Raising voice, risking retaliation: Events following interpersonal mistreatment in the workplace.* Journal of Occupational Health Psychology, 2003. **8**(4): 247.

Cozzi, D., and V. Vinel. *Risky, early, controversial: Puberty in medical discourses.* Social Science and Medicine, 2015. **143**:287–96.

Crawford, J. D., and D. C. Osler. *Body composition at menarche: The Frisch-Revelle hypothesis revisited.* Pediatrics, 1975. **56**(3): 449–58.

Crona Guterstam, Y., et al. *The cytokine profile of menstrual blood.* Acta Obstetricia et Gynecologica Scandinavica, 2021. **100**(2): 339–46.

Cryle, P., and E. Stephens. *Introduction.* In *Normality: A Critical Genealogy.* 2017: University of Chicago Press.

Cuda, A. *Is There a Link between COVID Vaccine and "Funky" Menstrual Periods? Experts Say It's Too Soon to Know.* Connecticut Post, April 19, 2021.

Czamanski-Cohen, J., et al. *Practice makes perfect: The effect of cognitive behavioral interventions during IVF treatments on women's perceived stress, plasma cortisol and pregnancy rates.* Journal of Reproductive Health and Medicine, 2016. **2**:S21–26.

Daggett, C. N. *The Birth of Energy: Fossil Fuels, Thermodynamics, and the Politics of Work.* Elements, ed. S. Alaimo and N. Starosielski. 2019: Duke University Press.

Dahvana Headley, M. Twitter post, March 2, 2020. https://twitter.com/MARIADAHVANA/status/1234588409357721600?s=20.

Daly, M., A. R. Sutin, and E. Robinson. *Perceived weight discrimination mediates the prospective association between obesity and physiological dysregulation: Evidence from a population-based cohort.* Psychological Science, 2019. **30**(7): 1030–39.

Daud, S., and A. A. Ewies. *Levonorgestrel-releasing intrauterine system: Why do some women dislike it?* Gynecological Endocrinology, 2008. **24**(12): 686–90.

Davis, J. L., and K. A. Smalls. *Dis/possession afoot: American (anthropological) traditions of anti-Blackness and coloniality.* Journal of Linguistic Anthropology, 2021. **31**(2): 275–82.

Dehlendorf, C., et al. *Recommendations for intrauterine contraception: A randomized trial of the effects of patients' race/ethnicity and socioeconomic status.* American Journal of Obstetrics and Gynecology, 2010. **203**(4): 319. e1–e8.

De Souza, M. J., and N. I. Williams. *Physiological aspects and clinical sequelae of energy deficiency and hypoestrogenism in exercising women.* Human Reproduction Update, 2004. **10**(5): 433–48.

Dhaliwal, A. *Woman World.* 2018: Drawn and Quarterly.

Dharmalingam, A. *The implications of menarche and wedding ceremonies for the status of women in a South Indian village.* Indian Anthropologist, 1994. **24**(1): 31–43.

Di Donato, N., et al. *Prevalence of adenomyosis in women undergoing surgery for endometriosis.* European Journal of Obstetrics and Gynecology and Reproductive Biology, 2014. **181**:289–93.

Dinh, T., et al. *Endocrinological effects of social exclusion and inclusion: Experimental evidence for adaptive regulation of female fecundity.* Hormones and Behavior, 2021. **130**:104934.

Downing, R. A., T. A. LaVeist, and H. E. Bullock. *Intersections of ethnicity and social class in provider advice regarding reproductive health.* American Journal of Public Health, 2007. **97**(10): 1803–7.

Dressler, W. W., K. S. Oths, and C. C. Gravlee. *Race and ethnicity in public health research: models to explain health disparities.* Annual Review of Anthropology, 2005. **34**:231–52.

Dube, S. R., et al. *The impact of adverse childhood experiences on health problems: Evidence from four birth cohorts dating back to 1900.* Preventive Medicine, 2003. **37**(3): 268–77.

Dubey, N., et al. *The ESC/E (Z) complex, an effector of response to ovarian steroids, manifests an intrinsic difference in cells from women with premenstrual dysphoric disorder.* Molecular Psychiatry, 2017. **22**(8): 1172–84.

DuBois, L. Z., et al. *Biocultural approaches to transgender and gender diverse experience and health: Integrating biomarkers and advancing gender/sex research.* American Journal of Human Biology, 2021. **33**(1): e23555.

Duboscq, J., et al. *The function of postconflict interactions: New prospects from the study of a tolerant species of primate.* Animal Behaviour, 2014. **87**:107–20.

Dubow, S. *Human origins, race typology and the other Raymond Dart.* African Studies, 1996. **55**(1): 1–30.

Duchesne, A., and J. C. Pruessner. *Association between subjective and cortisol stress response depends on the menstrual cycle phase.* Psychoneuroendocrinology, 2013. **38**(12): 3155–59.

Dufour, D. *The energetic cost of physical activity and the regulation of reproduction.* In *Reproduction and Adaptation: Topics in Human Reproductive Ecology*, ed. C. Maschie-Taylor and L. Rosetta. 2011: Cambridge University Press.

Edwards, V. J., et al. *Relationship between multiple forms of childhood maltreatment and adult mental health in community respondents: Results from the adverse childhood experiences study.* American Journal of Psychiatry, 2003. **160**(8): 1453–60.

Eisenlohr-Moul, T. A., et al. *Are there temporal subtypes of premenstrual dysphoric disorder? Using group-based trajectory modeling to identify individual differences in symptom change.* Psychological Medicine, 2020. **50**(6): 964–72.

Eisenlohr-Moul, T. A., et al. *Histories of abuse predict stronger within-person covariation of ovarian steroids and mood symptoms in women with menstrually related mood disorder.* Psychoneuroendocrinology, 2016. **67**:142–52.

Eisenlohr-Moul, T. A., et al. *Toward the reliable diagnosis of DSM-5 premenstrual dysphoric disorder: The Carolina Premenstrual Assessment Scoring System (C-PASS).* American Journal of Psychiatry, 2017. **174**(1): 51–59.

Ellington, J. E., et al. *Higher-quality human sperm in a sample selectively attach to oviduct (fallopian tube) epithelial cells in vitro.* Fertility and Sterility, 1999. **71**(5): 924–29.

Ellis, B. J., and M. J. Essex. *Family environments, adrenarche, and sexual maturation: A longitudinal test of a life history model.* Child Development, 2007. **78**(6): 1799–817.

Ellis, B. J., et al. *Quality of early family relationships and the timing and tempo of puberty: Effects depend on biological sensitivity to context.* Development and Psychopathology, 2011. **23**(1): 85.

Ellison, P. T. *On Fertile Ground.* 2001: Harvard University Press.

Ellison, P. T. *Skeletal growth, fatness, and menarcheal age—a comparison of 2 hypotheses.* Human Biology, 1982. **54**(2): 269–81.

Ellison, P. T., et al. *The ecological context of human ovarian function.* Human Reproduction, 1993. **8**(12): 2248–58.

Ellison, P. T., et al. *Moderate anxiety, whether acute or chronic, is not associated with ovarian suppression in healthy, well-nourished, Western women.* American Journal of Physical Anthropology, 2007. **134**(4): 513–19.

Elson, J. *Menarche, menstruation, and gender identity: Retrospective accounts from women who have undergone premenopausal hysterectomy.* Sex Roles, 2002. **46**(1–2): 37–48.

Embrick, D., and J. Williams. *Civility for Whom? Inside Higher Education,* November 16, 2018.

Evans, J., et al. *Menstrual fluid factors facilitate tissue repair: Identification and functional action in endometrial and skin repair.* FASEB Journal, 2019. **33**:000.

Evaristo, B. *Girl, Woman, Other.* 2019: Hamish Hamilton, an imprint of Penguin Random House UK.

Fairweather, D., S. Frisancho-Kiss, and N. R. Rose. *Sex differences in autoimmune disease from a pathological perspective.* American Journal of Pathology, 2008. **173**(3): 600–609.

Farland, L. V., et al. *Epidemiological and clinical risk factors for endometriosis.* In *Biomarkers for Endometriosis: State of the Art,* ed. T. D'Hooghe. 2019: Springer.

Farley, K. E., et al. *The association between contraceptive use at the time of conception and hypertensive disorders during pregnancy: A retrospective cohort study of PRAMS participants.* Maternal and Child Health Journal, 2014. **18**(8): 1779–85.

Fatima, S. *On the edge of knowing: Microaggression and epistemic uncertainty as a woman of color.* In *Surviving Sexism in Academia,* ed. K. Cole and H. Hassel. 2017: Routledge.

Fauconnier, A., and C. Chapron. *Endometriosis and pelvic pain: Epidemiological evidence of the relationship and implications.* Human Reproduction Update, 2005. **11**(6): 595–606.

Fausto-Sterling, A. *Gender/sex, sexual orientation, and identity are in the body: How did they get there?* Journal of Sex Research, 2019. **56**(4–5): 529–55.

Fausto-Sterling, A. *Sexing the Body: Gender Politics and the Construction of Sexuality.* 2000: Basic Books.

Felitti, V. J. *How Childhood Trauma Can Make You a Sick Adult.* Bigthink, video, 2019. https://bigthink.com/u/vincent-felitti.

Felitti, V. J., et al. *Relationship of childhood abuse and household dysfunction to many of the leading causes of death in adults: The Adverse Childhood Experiences (ACE) study.* American Journal of Preventive Medicine, 1998. **14**(4): 245–58.

Fielding-Singh, P., and A. Dmowska. *Obstetric gaslighting and the denial of mothers' realities.* Social Science and Medicine, 2022: 114938.

Finn, C. *Menstruation: A nonadaptive consequence of uterine evolution.* Quarterly Review of Biology, 1998. **73**(2): 163–73.

Firman, R. C., et al. *Postmating female control: 20 years of cryptic female choice.* Trends in Ecology and Evolution, 2017. **32**(5): 368–82.

Fitzwater, A. *Logistics. Clarkesworld*, no. 139, April 2018.

Forabosco, A., et al. *Morphometric study of the human neonatal ovary.* Anatomical Record, 1991. **231**(2): 201–8.

Freidenfelds, L. *The Modern Period: Menstruation in Twentieth-Century America.* 2009: Johns Hopkins University Press.

Freizinger, M., et al. *The prevalence of eating disorders in infertile women.* Fertility and Sterility, 2010. **93**(1): 72–78.

Frisch, R. E., and J. W. McArthur. *Menstrual cycles: Fatness as a determinant of minimum weight for height necessary for their maintenance or onset.* Science, 1974. **185**(4155): 949–51.

Fuller, E. A., et al. *Neuroimmune regulation of female reproduction in health and disease.* Current Opinion in Behavioral Sciences, 2019. **28**:8–13.

Futterman, M. *Another of Alberto Salazar's Runners Says He Ridiculed Her Body for Years. New York Times*, November 14, 2019.

Galton, F. *Inquiries into Human Faculty and Its Development.* 1883: Macmillan.

Geronimus, A. T. *Damned if you do: Culture, identity, privilege, and teenage childbearing in the United States.* Social Science and Medicine, 2003. **57**(5): 881–93.

Ghaderi, D., et al. *Sexual selection by female immunity against paternal antigens can fix loss of function alleles.* Proceedings of the National Academy of Sciences, 2011: **108**(43): 17443–48.

Gillies, V., et al. *Women's collective constructions of embodied practices through memory work: Cartesian dualism in memories of sweating and pain.* British Journal of Social Psychology, 2004. **43**(1): 99–112.

Goldman-Mellor, S., M. Hamer, and A. Steptoe. *Early-life stress and recurrent psychological distress over the lifecourse predict divergent cortisol reactivity patterns in adulthood.* Psychoneuroendocrinology, 2012. **37**(11): 1755–68.

Gomez, A. M., L. Fuentes, and A. Allina. *Women or LARC first? Reproductive autonomy and the promotion of long-acting reversible contraceptive methods.* Perspectives on Sexual and Reproductive Health, 2014. **46**(3): 171–75.

Gordy, C. *Anita Hill Defends Her Legacy. The Root*, October 18, 2011.

Gottlieb, A. *Rethinking female pollution: The Beng of Côte d'Ivoire.* Dialectical Anthropology, 1989. **14**(2): 65–79.

Gottschalk, M. S., et al. *The relation of number of childbirths with age at natural menopause: A population study of 310 147 women in Norway.* Human Reproduction, 2022. **37**(2): 333–40.

Gray, S. H., et al. *Salivary progesterone levels before menarche: A prospective study of adolescent girls.* Journal of Clinical Endocrinology and Metabolism, 2010. **95**(7): 3507–11.

Green, B. B., N. S. Weiss, and J. R. Daling. *Risk of ovulatory infertility in relation to body weight.* Fertility and Sterility, 1988. **50**(5): 721–26.

Green, E. *State-Mandated Mourning for Aborted Fetuses.* Atlantic, May 14, 2016.

Grzanka, P. R., and E. Schuch. *Reproductive anxiety and conditional agency at the intersections of privilege: A focus group study of emerging adults' perception of long-acting reversible contraception.* Journal of Social Issues, 2020. **76**(2): 270–313.

Guarner, F., et al. *Mechanisms of disease: The hygiene hypothesis revisited.* Nature Clinical Practice Gastroenterology and Hepatology, 2006. **3**(5): 275–84.

Gunnar, M., and K. Quevedo. *The neurobiology of stress and development.* Annual Review of Psychology, 2007. **58**(1): 145–73.

Gupta, S. K. *The human egg's zona pellucida.* Current Topics in Developmental Biology, 2018. **130**:379–411.

Halbreich, U., et al. *The prevalence, impairment, impact, and burden of premenstrual dysphoric disorder (PMS/PMDD).* Psychoneuroendocrinology, 2003. **28**:1–23.

Halbreich, U., and S. Karkun. *Cross-cultural and social diversity of prevalence of postpartum depression and depressive symptoms.* Journal of Affective Disorders, 2006. **91**(2–3): 97–111.

Halpern, V., et al. *Strategies to improve adherence and acceptability of hormonal methods of contraception.* Cochrane Database of Systematic Reviews, 2013. **26**(10): CD004317.

Hamburg, S. Twitter post, March 2, 2020. https://twitter.com/sarahrhamburg/status/1235227052778938368?s=20.

Harris, H. R., et al. *Early life abuse and risk of endometriosis.* Human Reproduction, 2018. **33**(9): 1657–68.

Harrison, F. V. *Decolonizing Anthropology: Moving Further toward an Anthropology for Liberation.* 1991: American Anthropological Association.

Harrison, F. V. *The persistent power of "race" in the cultural and political economy of racism.* Annual Review of Anthropology, 1995: 47–74.

Hasson, K. A. *Not a "real" period? Social and material constructions of menstruation.* Gender and Society, 2016. **30**(6): 958–83.

Healy, M., and W. Tagoe. *Dangerous blood: Menstruation, medicine and myth in early modern England.* In *National Healths: Gender, Sexuality and Health in a Cross-Cultural Context,* ed. M. Wooten and W. Tagoe, pp. 83–95. 2013: Routledge.

Henare, M. *Tapu, mana, mauri, hau, wairua: A Maori philosophy of vitalism and cosmos.* Indigenous Traditions and Ecology, 2001: 197–221.

Hennegan, J., and P. Montgomery. *Do menstrual hygiene management interventions improve education and psychosocial outcomes for women and girls in low and middle income countries? A systematic review.* PLOS ONE, 2016. **11**(2): e0146985.

Heuertz, R. M., et al. *Native and modified C-reactive protein bind different receptors on human neutrophils.* International Journal of Biochemistry and Cell Biology, 2005. **37**(2): 320–35.

Higgins, J. A. *Celebration meets caution: LARC's boons, potential busts, and the benefits of a repro-ductive justice approach.* Contraception, 2014. **89**(4): 237–41.

Hindson, B. *Attitudes towards menstruation and menstrual blood in Elizabethan England.* Journal of Social History, 2009: 89–114.

Hong, J. S., and D. L. Espelage. *A review of research on bullying and peer victimization in school: An ecological system analysis.* Aggression and Violent Behavior, 2012. **17**(4): 311–22.

Hopkinson, N. Twitter post, March 2, 2020. https://twitter.com/Nalo_Hopkinson/status/1234526087348015104?s=20.

Hosseini-Kamkar, N., C. Lowe, and J. B. Morton. *The differential calibration of the HPA axis as a function of trauma versus adversity: A systematic review and p-curve meta-analyses.* Neuroscience and Biobehavioral Reviews, 2021. **127**:154–330.

Houghton, L., et al. *A migrant study of pubertal timing and tempo in British-Bangladeshi girls at varying risk for breast cancer.* Breast Cancer Research, 2014. **16**:469.

Hrdy, S. B. *Mother Nature: A History of Mothers, Infants, and Natural Selection.* 1999: Pantheon.

Hrdy, S. B. *Mothers and others.* Natural History, 2001. **110**(4): 50–63.

Hrdy, S. B. *The Woman That Never Evolved.* 1981: Harvard University Press.

Huff, C. *In Texas, Abortion Laws Inhibit Care for Miscarriages.* NPR, May 10, 2022.

Huhmann, K. *Menses requires energy: A review of how disordered eating, excessive exercise, and high stress lead to menstrual irregularities.* Clinical Therapeutics, 2020. **42**(3): 401–7.

Hyde, J. S. *The gender similarities hypothesis.* American Psychologist, 2005. **60**(6): 581.

IJland, M. M., et al. *Endometrial wavelike movements during the menstrual cycle.* Fertility and Sterility, 1996. **65**(4): 746–49.

IJland, M. M., et al. *Relation between endometrial wavelike activity and fecundability in spontaneous cycles.* Fertility and Sterility, 1997. **67**(3): 492–96.

In Memoriam: Rose Epstein Frisch, Expert in Women's Fertility. T. H. Chan School of Public Health, Harvard University, 2015.

Insler, V., et al. *Sperm storage in the human cervix: A quantitative study.* Fertility and Sterility, 1980. **33**(3): 288–93.

Jabbour, H., et al. *Endocrine regulation of menstruation.* Endocrine Reviews, 2006. **27**(1): 17–46.

Jacobs, M. B., R. D. Boynton-Jarrett, and E. W. Harville. *Adverse childhood event experiences, fertility difficulties and menstrual cycle characteristics.* Journal of Psychosomatic Obstetrics and Gynecology, 2015. **36**(2): 46–57.

Jacobs-Huey, L. *The natives are gazing and talking back: Reviewing the problematics of positionality, voice, and accountability among "native" anthropologists.* American Anthropologist, 2002. **104**(3): 791–804.

Janik, E. *Marketplace of the Marvelous: The Strange Origins of Modern Medicine.* 2014: Beacon Press.

Janoowalla, H., et al. *The impact of menstrual hygiene management on adolescent health: The effect of Go! pads on rate of urinary tract infection in adolescent females in Kibogora, Rwanda.* International Journal of Gynecology and Obstetrics, 2020. **148**(1): 87–95.

Jasienska, G. *The Fragile Wisdom: An Evolutionary View on Women's Biology and Health.* 2013: Harvard University Press.

Jasienska, G., and P. T. Ellison. *Energetic factors and seasonal changes in ovarian function in women from rural Poland.* American Journal of Human Biology, 2004. **16**:563–80.

Jasienska, G., and P. T. Ellison. *Physical work causes suppression of ovarian function in women.* Proceedings of the Royal Society of London—Series B, 1998. **265**(1408): 1847–51.

Jasienska, G., et al. *Habitual physical activity and estradiol levels in women of reproductive age.* European Journal of Cancer Prevention, 2006. **15**:439–45.

Jasienska, G., and I. Thune. *Lifestyle, hormones, and risk of breast cancer.* British Medical Journal, 2001. **322**:586–87.

Jia, M., K. Dahlman-Wright, and J.-Å. Gustafsson. *Estrogen receptor alpha and beta in health and disease.* Best Practice and Research Clinical Endocrinology and Metabolism, 2015. **29**(4): 557–68.

Joel, D. *Genetic-gonadal-genitals sex (3G-sex) and the misconception of brain and gender, or, why 3G-males and 3G-females have intersex brain and intersex gender.* Biology of Sex Differences, 2012. **3**(1): 27.

Johnson, M. E. *Menstrual justice.* UC Davis Law Review, 2019. **53**:1.

Johnson, P., Widnall, S. E., and Benya, F. F., eds. *Sexual Harassment of Women: Climate, Culture, and Consequences in Academic Sciences, Engineering, and Medicine.* 2018: National Academies Press.

Jones, C. E. *The pain of endo existence: Toward a feminist disability studies reading of endometriosis.* Hypatia, 2016. **31**(3): 554–71.

Jones, K., and T. Okun. *White supremacy culture.* In *Dismantling Racism: A Workbook for Social Change Groups.* 2001: Dismantlingracism.org.

Junkins, E., K. Lee, and K. Clancy. *Enhancing Knowledge through Engagement: Participation in an Unpaid Survey-Based Health Research.* International Congress of Qualitative Inquiry, virtual online meeting, May 2022.

Katsu, N., K. Yamada, and M. Nakamichi. *Functions of post-conflict affiliation with a bystander differ between aggressors and victims in Japanese macaques.* Ethology, 2018. **124**(2): 94–104.

Kessler, S. J. *The medical construction of gender: Case management of intersexed infants.* Signs: Journal of Women in Culture and Society, 1990. **16**(1): 3–26.

King, S. *Carrie.* 1974: Doubleday.

King, S. *The Stand.* 1978: Doubleday.

Klausen, S., and A. Bashford. *Fertility control: Eugenics, neo-Malthusianism, and feminism.* In *The Oxford Handbook of the History of Eugenics*, pp. 98–115. 2010: Oxford University Press.

Kramer, K. L. *Early sexual maturity among Pume foragers of Venezuela: Fitness implications of teen motherhood.* American Journal of Physical Anthropology, 2008. **136**(3): 338–50.

Lady with Desire to Run Crashed Marathon. New York Times, April 23, 1967.

Landau, M. *Narratives of Human Evolution.* 1991: Yale University Press.

Latthe, P., et al. *WHO systematic review of prevalence of chronic pelvic pain: A neglected reproductive health morbidity.* BMC Public Health, 2006. **6**(1): 177.

Layne, L. L. *Introduction.* In *Feminist Technology*, ed. L. L. Layne, S. Vostral, and K. Boyer. 2010: University of Illinois Press.

Lee, J. A., and C. J. Pausé. *Stigma in practice: Barriers to health for fat women.* Frontiers in Psychology, 2016. 7:2063.

Lee, K. M. N., et al. *Bone density and frame size in adult women: Effects of body size, habitual use, and life history.* American Journal of Human Biology, 2020. **33**(2): e23502.

Lee, K. M. N., et al. *Investigating trends in those who experience menstrual bleeding changes after SARS-CoV-2 vaccination.* medRxiv, 2022: 2021.10.11.21264863.

Lee, K. M. N., et al., *Investigating trends in those who experience menstrual bleeding changes after SARS-CoV-2 vaccination*, 2022: Science Advances, 8(29) DOI: 10.1126/sciadv.abm7201.

Lee, K. M. N., et al. *Measuring menstruation: Methodological difficulties in studying things we don't talk about* (poster presented at the Human Biology Association). American Journal of Human Biology, 2022. **34**(S2): e23740.

Lee, K. M. N., et al. *Physical activity in women of reproductive age in a transitioning rural Polish population.* American Journal of Human Biology, 2019. **31**(3): e23231.

Lee, K. M. N., E. Junkins, and K. Clancy. *Menstrual experiences after COVID-19 vaccination in a non-menstruating gender diverse sample* (poster presentation). Experimental Biology, 2022. **36**(S1). https://doi.org/10.1096/fasebj.2022.36.S1.R5912.

Legro, R. S. *Effects of obesity treatment on female reproduction: Results do not match expectations.* Fertility and Sterility, 2017. **107**(4): 860–67.

Leidy Sievert, L. *Should women menstruate? An evolutionary perspective on menstrual-suppressing oral contraceptives.* In *Evolutionary Medicine and Health: New Perspectives*, ed. W. Trevathan, E. O. Smith, and J. J. McKenna, pp. 181–95. 2008: Oxford University Press.

Lester, C. *How Childhood Trauma Affects Health* (blog), July 13, 2017. http://blogs.wgbh.org/innovation-hub/2017/7/13/felitti-ace/.

Levenson, E., H. Yan, and E. Wolfe. *Uvalde Mass Shooter Was Not Confronted by Police before He Entered the school, Texas Official Says.* CNN, May 14, 2022.

Lewis, J. A., et al. *Racial microaggressions and sense of belonging at a historically white university.* American Behavioral Scientist, 2019. **65**(8): 1049–71.

Liboiron, M. *Introduction.* In *Pollution Is Colonialism.* 2021: Duke University Press.

Liboiron, M. *Pollution Is Colonialism.* 2021: Duke University Press.

Liboiron, M., M. Tironi, and N. Calvillo. *Toxic politics: Acting in a permanently polluted world.* Social Studies of Science, 2018. **48**(3): 331–49.

Lilly, S. *Can COVID-19 Vaccine Impact Your Menstrual Cycle? Doctors Address Side-Effects Concerns. Dr. Viray: "That's Just Not How These Vaccines Work."* 6 News Richmond, April 14, 2021.

Linna, M. S., et al. *Reproductive health outcomes in eating disorders.* International Journal of Eating Disorders, 2013. **46**(8): 826–33.

Loman, M. M., and M. R. Gunnar. *Early experience and the development of stress reactivity and regulation in children.* Neuroscience and Biobehavioral Reviews, 2010. **34**(6): 867–76.

Luna, Z., and K. Luker. *Reproductive justice.* Annual Review of Law and Social Science, 2013. **9**:327–52.

Maack, D. J., E. Buchanan, and J. Young. *Development and psychometric investigation of an inventory to assess fight, flight, and freeze tendencies: The fight, flight, freeze questionnaire.* Cognitive Behaviour Therapy, 2015. **44**(2): 117–27.

Macht, D. I., and D. S. Lubin. *A phyto-pharmacological study of menstrual toxin.* Journal of Pharmacology and Experimental Therapeutics, 1923. **22**(5): 413–66.

Mack, N., et al. *Strategies to improve adherence and continuation of shorter-term hormonal methods of contraception.* Cochrane Database of Systematic Reviews, 2019. **4**(4): CD004317.

Maia, J., et al. *The endocannabinoid system expression in the female reproductive tract is modulated by estrogen.* Journal of Steroid Biochemistry and Molecular Biology, 2017. **174**:40–47.

Major, B., W. J. Quinton, and S. K. McCoy. *Antecedents and consequences of attributions to discrimination: Theoretical and empirical advances.* Advances in Experimental Social Psychology, 2002. **34**:251–330.

Major, B., A. J. Tomiyama, and J. M. Hunger. *The Negative and Bidirectional Effects of Weight Stigma on Health.* 2018: Oxford Handbooks Online.

Male, V. *Are COVID-19 vaccines safe in pregnancy?* Nature Reviews Immunology, 2021. **21**(4): 200–201.

Mann, E. S. *The power of persuasion: Normative accountability and clinicians' practices of contraceptive counseling.* SSM-Qualitative Research in Health, 2022. **2**:100049.

Marcus, M. D., T. L. Loucks, and S. L. Berga. *Psychological correlates of functional hypothalamic amenorrhea.* Fertility and Sterility, 2001. **76**(2): 310–16.

Martin, E. *The egg and the sperm: How science has constructed a romance based on stereotypical male-female roles.* Signs: Journal of Women in Culture and Society, 1991. **16**(3): 485–501.

Martin, E. *The Woman in the Body: A Cultural Analysis of Reproduction.* 1980: Beacon Press.

Martinez, P. E., et al. *5 α-reductase inhibition prevents the luteal phase increase in plasma allopregnanolone levels and mitigates symptoms in women with premenstrual dysphoric disorder.* Neuropsychopharmacology, 2016. **41**(4): 1093–102.

Mather, V. *Cain's Abuse Allegations against Salazar Cause More Upheaval in Track World. New York Times,* November 8, 2019.

Matheson, K., and H. Anisman. *Anger and shame elicited by discrimination: Moderating role of coping on action endorsements and salivary cortisol.* European Journal of Social Psychology, 2009. **39**(2): 163–85.

Matsumoto, T., et al. *Comparison between retrospective premenstrual symptoms and prospective late-luteal symptoms among college students.* Gynecological and Reproductive Endocrinology and Metabolism, 2021. **2**(1): 31–41.

Maybin, J. A., and H. O. Critchley. *Menstrual physiology: Implications for endometrial pathology and beyond.* Human Reproduction Update, 2015. **21**(6): 748–61.

McKinley, R. *The Blue Sword.* 1982: Greenwillow Books.

McKittrick, K. *The smallest cell remembers a sound.* In *Dear Science and Other Stories.* 2021: Duke University Press.

McPherson, M. E., and L. Korfine. *Menstruation across time: Menarche, menstrual attitudes, experiences, and behaviors.* Womens Health Issues, 2004. **14**(6): 193–200.

Meltzer-Brody, S., et al. *Trauma and posttraumatic stress disorder in women with chronic pelvic pain.* Obstetrics and Gynecology, 2007. **109**(4): 902–8.

Mendenhall, R. *The medicalization of poverty in the lives of low-income Black mothers and children.* Journal of Law, Medicine and Ethics, 2018. **46**(3): 644–50.

Merriam-Webster's Collegiate Dictionary. Vol. 2. 2004: Merriam-Webster.

Metsios, G. S., R. H. Moe, and G. D. Kitas. *Exercise and inflammation.* Best Practice and Research Clinical Rheumatology, 2020. **34**(2): 101504.

Mihm, M., and A. Evans. *Mechanisms for dominant follicle selection in monovulatory species: A comparison of morphological, endocrine and intraovarian events in cows, mares and women.* Reproduction in Domestic Animals, 2008. **43**:48–56.

Mills, C. W. *The Racial Contract*. 1997: Cornell University Press.

Moore, A. M. *Victorian medicine was not responsible for repressing the clitoris: Rethinking homology in the long history of women's genital anatomy*. Signs: Journal of Women in Culture and Society, 2018. **44**(1): 53–81.

Morgenroth, T., and M. K. Ryan. *The effects of gender trouble: An integrative theoretical framework of the perpetuation and disruption of the gender/sex binary*. Perspectives on Psychological Science, 2020. **16**(6): 1113–42.

Morrison, T., A. Dinno, and T. Salmon. *The erasure of intersex, transgender, nonbinary, and agender experiences through misuse of sex and gender in health research*. American Journal of Epidemiology, 2021. **190**(12): 2712–17.

Mountjoy, M., et al. *The IOC consensus statement: Beyond the female athlete triad—relative energy deficiency in sport (RED-S)*. British Journal of Sports Medicine, 2014. **48**(7): 491–97.

Muller, M. N., and R. W. Wrangham. *Sexual Coercion in Primates and Humans*. 2009: Harvard University Press.

Mullings, L. *Interrogating racism: Toward an antiracist anthropology*. Annual Review of Anthropology, 2005. **34**:667–93.

Murdock, M. *The Heroine's Journey: Woman's Quest for Wholeness*. 30th anniversary ed. 2020: Shambhala. Originally printed in 1990.

Murray, M. *Race-ing Roe: Reproductive justice, racial justice, and the battle for Roe v. Wade*. Harvard Law Review, 2020. **134**:2025.

Nadal, K. L., et al. *Sexual orientation and transgender microaggressions*. In *Microaggressions and Marginality: Manifestation, Dynamics, and Impact*, ed. Derald Wing Sue, pp. 217–40. 2010: Wiley.

Nadeau, J. H. *Do gametes woo? Evidence for their nonrandom union at fertilization*. Genetics, 2017. **207**(2): 369–87.

Nagel, T. *The View from Nowhere*. 1989: Oxford University Press.

Natt, A. M., F. Khalid, and S. S. Sial. *Relationship between examination stress and menstrual irregularities among medical students of Rawalpindi Medical University*. Journal of Rawalpindi Medical College, 2018. **22**(S–1): 44–47.

Negriff, S. *ACEs are not equal: Examining the relative impact of household dysfunction versus childhood maltreatment on mental health in adolescence*. Social Science and Medicine, 2020. **245**:112696.

Nelson, R., et al. *Signaling safety: Characterizing fieldwork experiences and their implications for career trajectories*. American Anthropologist, 2017. **119**(4): 710–22.

Nissen, A. *Transgression, pollution, deformity, bewitchment: Menstruation and supernatural threat in late medieval and early modern England, 1250–1750*. PhD diss., State University of New York Empire State College, 2017.

Nixon, B., R. J. Aitken, and E. A. McLaughlin. *New insights into the molecular mechanisms of sperm-egg interaction*. Cellular and Molecular Life Sciences, 2007. **64**(14): 1805–23.

Norgan, N. *The beneficial effects of body fat and adipose tissue in humans*. International Journal of Obesity, 1997. **21**(9): 738–46.

Norman, A. *Ask Me about My Uterus: A Quest to Make Doctors Believe in Women's Pain*. 2018: Bold Type Books.

Norman, R., S. Mahabeer, and S. Masters. *Ethnic differences in insulin and glucose response to glucose between white and Indian women with polycystic ovary syndrome*. Fertility and Sterility, 1995. **63**(1): 58–62.

Novak, M. *Reckless Breeding of the Unfit: Earnest Hooton, Eugenics and the Human Body of the Year 2000. Smithsonian* magazine, February 12, 2013.

O'Brien, M. H. *Being a scientist means taking sides.* Professional Biologist, 1993. **43**(10): 706–8.

Odwe, G., et al. *Method-specific beliefs and subsequent contraceptive method choice: Results from a longitudinal study in urban and rural Kenya.* PLOS ONE, 2021. **16**(6): e0252977.

Oreskes, N. *Science awry.* In *Why Trust Science?*, ed. N. Oreskes, pp. 69–146. 2019: Princeton University Press.

Oreskes, N. *Why trust science? Perspectives from the history and philosophy of science.* In *Why Trust Science?*, ed. N. Oreskes, pp. 15–68. 2019: Princeton University Press.

O'Rourke, M. T., and P. T. Ellison. *Age-related patterns of salivary estradiol across the menstrual-cycle.* American Journal of Physical Anthropology, 1989. **78**(2): 281.

Ortega, F. B., et al. *The Fat but Fit Paradox: What We Know and Don't Know about It.* 2018: BMJ and British Association of Sport and Exercise Medicine.

Oruko, K., et al. *"He is the one who is providing you with everything so whatever he says is what you do": A qualitative study on factors affecting secondary schoolgirls' dropout in rural Western Kenya.* PLOS ONE, 2015. **10**(12): e0144321.

Owens, D. C. *Medical Bondage: Race, Gender, and the Origins of American Gynecology.* 2017: University of Georgia Press.

Palm-Fischbacher, S., and U. Ehlert. *Dispositional resilience as a moderator of the relationship between chronic stress and irregular menstrual cycle.* Journal of Psychosomatic Obstetrics and Gynecology, 2014. **35**(2): 42–50.

Park, S., et al. *Impact of childhood exposure to psychological trauma on the risk of psychiatric disorders and somatic discomfort: Single vs. multiple types of psychological trauma.* Psychiatry Research, 2014. **219**(3): 443–49.

Park, S. U., L. Walsh, and K. M. Berkowitz. *Mechanisms of ovarian aging.* Reproduction, 2021. **162**(2): R19–R33.

Parker, R., T. Larkin, and J. Cockburn. *A visual analysis of gender bias in contemporary anatomy textbooks.* Social Science and Medicine, 2017. **180**:106–13

Pascal, M., et al. *Microbiome and allergic diseases.* Frontiers in Immunology, 2018. **9**:1584.

Patterson, R., et al. *Sedentary behaviour and risk of all-cause, cardiovascular and cancer mortality, and incident type 2 diabetes: A systematic review and dose response meta-analysis.* European Journal of Epidemiology, 2018. **33**(9): 811–29.

Pauli, S. A., and S. L. Berga. *Athletic amenorrhea: Energy deficit or psychogenic challenge?* Annals of the New York Academy of Sciences, 2010. **1205**(1): 33–38.

Payne, J. L., and J. Maguire. *Pathophysiological mechanisms implicated in postpartum depression.* Frontiers in Neuroendocrinology, 2019. **52**:165–80.

Perez-Lopez, F. R., et al. *Uterine or paracervical lidocaine application for pain control during intrauterine contraceptive device insertion: A meta-analysis of randomised controlled trials.* European Journal of Contraception and Reproductive Health Care, 2018. **23**(3): 207–17.

Perlstein, M., and A. Matheson. *Allergy due to menotoxin of pregnancy.* Archives of Pediatrics and Adolescent Medicine, 1936. **52**(2): 303.

Petersen, A. M. W., and B. K. Pedersen. *The anti-inflammatory effect of exercise.* Journal of Applied Physiology, 2005. **98**(4): 1154–62.

Phillips-Howard, P. A., et al. *Menstrual cups and sanitary pads to reduce school attrition, and sexually transmitted and reproductive tract infections: A cluster randomised controlled feasibility study in rural western Kenya.* BMJ Open, 2016. **6**(11): e013229.

Pickles, V. *Prostaglandins and dysmenorrhea: Historical survey.* Acta Obstetricia Gynecologica Scandinavica suppl., 1979. **87**:7–12.

Piepzna-Samarasinha, L. L. *Crip emotional intelligence.* In *Care Work: Dreaming Disability Justice.* 2018: Arsenal Pulp Press.

Piepzna-Samarasinha, L. L. *Making space accessible is an act of love for our communities.* In *Care Work: Dreaming Disability Justice.* 2018: Arsenal Pulp Press.

Pilar Matud, M., J. M. Bethencourt, and I. Ibáñez. *Relevance of gender roles in life satisfaction in adult people.* Personality and Individual Differences, 2014. **70**:206–11.

Pilver, C. E., et al. *Exposure to American culture is associated with premenstrual dysphoric disorder among ethnic minority women.* Journal of Affective Disorders, 2011. **130**(1–2): 334–41.

Pilver, C. E., et al. *Lifetime discrimination associated with greater likelihood of premenstrual dysphoric disorder.* Journal of Women's Health, 2011. **20**(6): 923–31.

Pontzer, H., et al. *Energy expenditure and activity among Hadza hunter-gatherers.* American Journal of Human Biology, 2015. **27**(5): 628–37.

Preiss, D. *A Young Woman Died in a Menstrual Hut on Nepal.* NPR, November 28, 2016.

Prescod-Weinstein, C. *Ain't I a woman? At the intersection of gender, race, and sexuality.* In *Women in Astronomy* (blog). 2014.

Pretty, C., et al. *Adverse childhood experiences and the cardiovascular health of children: A cross-sectional study.* BMC Pediatrics, 2013. **13**(1): 1–8.

Prior, J. C. *Luteal phase defects and anovulation: Adaptive alterations occurring with conditioning exercise.* Seminars in Reproductive Endocrinology, 1985. **3**:27–33.

Prior, J. C. *Perimenopause: The complex endocrinology of the menopausal transition.* Endocrine Reviews, 1998. **19**(4): 397–428.

Prior, J. C., et al. *Ovulation prevalence in women with spontaneous normal-length menstrual cycles—a population-based cohort from HUNT3, Norway.* PLOS ONE, 2015. **10**(8): e0134473.

Prior, J. C., et al. *Reversible luteal phase changes and infertility associated with marathon training.* Lancet, 1982. **320**(8292): 269–70.

Prior, J. C., and C. L. Hitchcock. *The endocrinology of perimenopause: Need for a paradigm shift.* Frontiers in Bioscience, 2011. **3**:474–86.

Profet, M. *Menstruation as a defense against pathogens transported by sperm.* Quarterly Review of Biology, 1993. **68**(3): 335–86.

Promislow, J. H., et al. *Bleeding following pregnancy loss before 6 weeks' gestation.* Human Reproduction, 2007. **22**(3): 853–57.

Przybylo, E., and B. Fahs. *Feels and flows: On the realness of menstrual pain and cripping menstrual chronicity.* Feminist Formations, 2018. **30**(1): 206–29.

Rachlin, K., G. Hansbury, and S. T. Pardo. *Hysterectomy and oophorectomy experiences of female-to-male transgender individuals.* International Journal of Transgenderism, 2010. **12**(3): 155–66.

Rebuelta-Cho, A. P. *"Give her the baby's hat so she can bite it": Obstetric violence in Flores, Indonesia.* Moussons-Recherche En Sciences Humaines Sur L Asie Du Sud-Est, 2021. (38): 57–84.

Reed, J. L., et al. *Energy availability discriminates clinical menstrual status in exercising women.* Journal of the International Society of Sports Nutrition, 2015. **12**(1): 11.

Reiches, M. *Reproductive justice and the history of prenatal supplementation: Ethics, birth spacing, and the "priority infant" model in the Gambia: Winner of the 2019 Catharine Stimpson Prize for Outstanding Feminist Scholarship.* Signs: Journal of Women in Culture and Society, 2019. **45**(1): 3–26.

Reid, H. *Letter: The brass-ring sign.* Lancet, 1974. **1**(7864): 988.

Reid, R. L., and C. N. Soares. *Premenstrual dysphoric disorder: Contemporary diagnosis and management.* Journal of Obstetrics and Gynaecology Canada, 2018. **40**(2): 215–23.

Richey, C. R., et al. *Gender and sexual minorities in astronomy and planetary science face increased risks of harassment and assault.* Bulletin of the American Astronomical Society, 2019. **51**(4). https://doi.org/10.3847/25c2cfeb.c985281e.

Riddle, J. M. *Birth control (abortion and contraception).* In *The Encyclopedia of Ancient History*, pp. 1–2. 2013: Wiley-Blackwell.

Rizk, B., et al. *Recurrence of endometriosis after hysterectomy.* Facts, Views and Vision in ObGyn, 2014. **6**(4): 219.

Roberts, T. A., et al. *"Feminine protection": The effects of menstruation on attitudes toward women.* Psychology of Women Quarterly, 2002. **26**:131–39.

Roberts, T. A., and S. Hansen. *Association of hormonal contraception with depression in the postpartum period.* Contraception, 2017. **96**(6): 446–52.

Robertson, D. M., et al. *Random start or emergency IVF/in vitro maturation: A new rapid approach to fertility preservation.* Women's Health, 2016. **12**(3): 339–49.

Robinson, W. R., et al. *For U.S. Black women, shift of hysterectomy to outpatient settings may have lagged behind white women: A claims-based analysis, 2011–2013.* BMC Health Services Research, 2018. **17**(526): 1–9.

Rocha Beardall, T. *Settler simultaneity and anti-Indigenous racism at land-grant universities.* Sociology of Race and Ethnicity, 2022. **8**(1): 197–212.

Rodnite Lemay, H. *Women's Secrets: A Translation of Pseudo-Albertus Magnus' de Secretis Mulierum with Commentaries.* 1992: State University of New York Press.

Rodrigues, M. A. *Emergence of sex-segregated behavior and association patterns in juvenile spider monkeys.* Neotropical Primates, 2014. **21**(2): 183–88.

Rodrigues, M. A. *Stress and sociality in a patrilocal primate: Do female spider monkeys tend-and-befriend?* PhD diss., Ohio State University, 2013.

Rodrigues, M. A., and E. R. Boeving. *Comparative social grooming networks in captive chimpanzees and bonobos.* Primates, 2019. **60**(3): 191–202.

Rodrigues, M. A., et al. *From maternal tending to adolescent befriending: The adolescent transition of social support.* American Journal of Primatology, 2019. **82**(11): e23050.

Rodrigues, M. A., R. Mendenhall, and K. Clancy. *"There's realizing, and then there's realizing": How social support can counter gaslighting of women of color scientists.* Journal of Women and Minorities in Science and Engineering, 2021. **27**(2): 1–23.

Rogers, M. P. *Variation in age at menarche and reproductive function in the United States and Poland.* PhD diss., University of Illinois, 2018.

Rogers, M. P., et al. *Declining ages at menarche in an agrarian rural region of Poland.* American Journal of Human Biology, 2020. **32**(3): e23362.

Rogers, M. P., et al. *The effects of adolescent-parent open communication on age at menarche.* Forthcoming.

Rogers-LaVanne, M. P. *Variation in age at menarche and adult reproductive function: The role of energetic and psychosocial stressors.* PhD diss., University of Illinois Urbana-Champaign, 2018.

Rogers-LaVanne, M. P., and K. Clancy. *Menstruation: Causes, consequences, and context.* In *Routledge Handbook of Anthropology and Reproduction,* ed. S. Han and C. Tomori. 2021: Routledge.

Rose, T. *When U.S. Air Force Discovered the Flaw of Averages.* Star (Toronto), January 16, 2016.

Roved, J., H. Westerdahl, and D. Hasselquist. *Sex differences in immune responses: Hormonal effects, antagonistic selection, and evolutionary consequences.* Hormones and Behavior, 2017. **88**:95–105.

Ryu, A., and T.-H. Kim. *Premenstrual syndrome: A mini review.* Maturitas, 2015. **82**(4): 436–40.

Sanabria, E. *Introduction.* In *Plastic Bodies: Sex Hormones and Menstrual Suppression in Brazil.* 2016: Duke University Press.

Sanders, K. M., et al. *Heightened cortisol response to exercise challenge in women with functional hypothalamic amenorrhea.* American Journal of Obstetrics and Gynecology, 2018. **218**(2): 230. e1–e6.

Schalk, S. *The future of bodyminds, bodyminds of the future.* In *Bodyminds Reimagined: (Dis)ability, Race, and Gender in Black Women's Speculative Fiction.* 2018: Duke University Press.

Schalk, S. *Introduction.* In *Bodyminds Reimagined: (Dis)ability, Race, and Gender in Black Women's Speculative Fiction.* 2018: Duke University Press.

Schiebinger, L. *Primatology, archaeology, and human origins—feminist interventions.* In *Conference on Equal Rites, Unequal Outcomes—Women in American Research Universities.* 1998. Kluwer Academic/Plenum.

Schoenaker, D. A., et al. *Socioeconomic position, lifestyle factors and age at natural menopause: A systematic review and meta-analyses of studies across six continents.* International Journal of Epidemiology, 2014. **43**(5): 1542–62.

Schuller, K. *The Biopolitics of Feeling.* 2018: Duke University Press.

Schurr, C. *From biopolitics to bioeconomies: The ART of (re-) producing white futures in Mexico's surrogacy market.* Environment and Planning D: Society and Space, 2017. **35**(2): 241–62.

Sear, R., R. Mace, and I. A. McGregor. *Maternal grandmothers improve nutritional status and survival of children in rural Gambia.* Proceedings of the Royal Society of London Series B—Biological Sciences, 2000. **267**(1453): 1641–47.

Sear, R., P. Sheppard, and D. A. Coall. *Cross-cultural evidence does not support universal acceleration of puberty in father-absent households.* Philosophical Transactions of the Royal Society B—Biological Sciences, 2019. **374**(1770): 20180124.

Sebring, J. C. *Towards a sociological understanding of medical gaslighting in western health care.* Sociology of Health and Illness, 2021. **43**(9): 1951–64.

Segers, G. *Here Are Some of the Questions Anita Hill Answered in 1991.* CBS News, September 19, 2018.

Seller, M. S. *Dr. Clarke vs. "the Ladies": Coeducation and Women's Roles in the 1870's.* Paper presented at the Annual Meeting of the American Educational Research Association, Montreal, 1983.

Serin, I. S., et al. *Effects of hypertension and obesity on endometrial thickness.* European Journal of Obstetrics and Gynecology and Reproductive Biology, 2003. **109**:72–75.

Shah, R., and D. C. Newcomb. *Sex bias in asthma prevalence and pathogenesis.* Frontiers in Immunology, 2018. **9**:2997.

Shapin, S. *Never Pure: Historical Studies of Science as If It Was Produced by People with Bodies, Situated in Time, Space, Culture, and Society, and Struggling for Credibility and Authority.* 2010: Johns Hopkins University Press.

Shapin, S. *Placing the view from nowhere: Historical and sociological problems in the location of science.* Transactions of the Institute of British Geographers, 1998. **23**(1): 5–12.

Shapiro, H. L. *A portrait of the American people.* Natural History, 1945. **54**:248–55.

Sharma, B., and K. Schultz. *Woman and 2 Children Die in Nepal Menstruation Hut.* New York Times, January 9, 2019.

Shelley, M. W. *Frankenstein.* 1968: Golden Press.

Shires, D. A., and K. Jaffee. *Factors associated with health care discrimination experiences among a national sample of female-to-male transgender individuals.* Health and Social Work, 2015. **40**(2): 134–41.

Shotwell, A. *Against Purity: Living Ethically in Compromised Times.* 2016: University of Minnesota Press.

Sievert, L. L. *Menopause across cultures: Clinical considerations.* Menopause, 2014. **21**(4): 421–23.

Singer, M. R., et al. *Pediatricians' knowledge, attitudes and practices surrounding menstruation and feminine products.* International Journal of Adolescent Medicine and Health, 2020. **34**(3): n.p.

Skloot, R. *The Immortal Life of Henrietta Lacks.* 2011: Crown.

Slootmaker, S. M., et al. *Disagreement in physical activity assessed by accelerometer and self-report in subgroups of age, gender, education and weight status.* International Journal of Behavioral Nutrition and Physical Activity, 2009. **6**(1): 1–10.

Sniekers, M. *From little girl to young woman: The menarche ceremony in Fiji.* Fijian Studies, 2005. **3**(2): 397.

Sokoloff, N. C., M. Misra, and K. E. Ackerman. *Exercise, training, and the hypothalamic-pituitary-gonadal axis in men and women.* Sports Endocrinology, 2016. **47**:27–43.

Sproston, N. R., and J. J. Ashworth. *Role of C-reactive protein at sites of inflammation and infection.* Frontiers in Immunology, 2018. **9**:754.

Spruck Wrigley, S. *Space Travel Loses Its Allure When You've Lost Your Moon Cup,* Flash Fiction Online, October 2015.

Steele, H., et al. *Adverse childhood experiences, poverty, and parenting stress.* Canadian Journal of Behavioural Science/Revue Canadienne des sciences du comportement, 2016. **48**(1): 32.

Steele, J. R., and N. Ambady. *"Math is hard!" The effect of gender priming on women's attitudes.* Journal of Experimental Social Psychology, 2006. **42**(4): 428–36.

Steinfeldt, J. A., et al. *Muscularity beliefs of female college student-athletes.* Sex Roles, 2011. **64**(7–8): 543–54.

Stephens, E. *The Object of Normality: The "Search for Norma" Competition.* Queer Objects Symposium, Gender Institute and School of Literature, Languages and Linguistics, Australian National University, Canberra, 2014.

Stewart, E. A., L. T. Shuster, and W. A. Rocca. *Reassessing hysterectomy.* Minnesota Medicine, 2012. **95**(3): 36.

Stock, N. *Some People Are Reporting Abnormal Periods after a COVID-19 Vaccine. U. of I. Professor Is Looking for Answers. Chicago Tribune*, April 20, 2021.

Strasser, J., et al. *Access to removal of long-acting reversible contraceptive methods is an essential component of high-quality contraceptive care.* Women's Health Issues, 2017. **27**(3): 253–55.

Strassmann, B. I. *The biology of menstruation in Homo sapiens: Total lifetime menses, fecundity, and nonsynchrony in a natural-fertility population.* Current Anthropology, 1997. **38**(1): 123–29.

Strassmann, B. I. *Energy economy in the evolution of menstruation.* Evolutionary Anthropology, 1996. **5**(5): 157–64.

Strassmann, B. I. *The evolution of endometrial cycles and menstruation.* Quarterly Review of Biology, 1996. **71**(2): 181–220.

Strings, S. *Birth of the ascetic aesthetic.* In *Fearing the Black Body: The Racial Origins of Fat Phobia.* 2019: New York University Press.

Strings, S. *Fearing the Black Body: The Racial Origins of Fat Phobia.* 2019: New York University Press.

Strings, S. *The rise of the big Black woman.* In *Fearing the Black Body: The Racial Origins of Fat Phobia.* 2019: New York University Press.

Strout, E. *That's Not Fat: How Ryan Hall Gained 40 Pounds of Muscle. Runner's World*, May 3, 2016.

Stubblefield, A. *"Beyond the pale": Tainted whiteness, cognitive disability, and eugenic sterilization.* Hypatia, 2007. **22**(2): 162–81.

Subar, A. F., et al. *Addressing current criticism regarding the value of self-report dietary data.* Journal of Nutrition, 2015. **145**(12): 2639–45.

Sugrue, T. *White America's Age-Old, Misguided Obsession with Civility. New York Times*, June 29, 2018, A23.

Suh, B., et al. *Hypercortisolism in patients with functional hypothalamic-amenorrhea.* Journal of Clinical Endocrinology and Metabolism, 1988. **66**(4): 733–39.

Swehla, T. Twitter post, March 2, 2020. https://twitter.com/SwehlaTessa/status/1234532953675509760?s=20.

Tarawneh, O., and H. Tarawneh. *Immune thrombocytopenia in a 22-year-old post Covid-19 vaccine.* American Journal of Hematology, 2021. **96**(5): E133–34.

Taylor, A. H., et al. *Histomorphometric evaluation of cannabinoid receptor and anandamide modulating enzyme expression in the human endometrium through the menstrual cycle.* Histochemistry and Cell Biology, 2010. **133**(5): 557–65.

Taylor, S. E. *Tend and befriend: Biobehavioral bases of affiliation under stress.* Current Directions in Psychological Science, 2006. **15**(6): 273–77.

Tehrani, F. R., et al. *The prevalence of polycystic ovary syndrome in a community sample of Iranian population: Iranian PCOS prevalence study.* Reproductive Biology and Endocrinology, 2011. **9**(1): 39.

Theobald, B. *To instill the hospital habit.* In *Reproduction on the Reservation: Pregnancy, Childbirth, and Colonialism in the Long Twentieth Century*, pp. 44–70. 2019: University of North Carolina Press.

Thometz, K. *Do COVID-19 Vaccines Impact Menstrual Cycles?.* WTTW News (Chicago), May 7, 2021.

Thompson, C., et al. *10 Dead in Buffalo Supermarket Attack Police Call Hate Crime.* AP News, May 14, 2022.

Thompson, T., ed. *The Murders of Molly Southbourne*. 2017: Tor.

Thompson, T., ed. *The Survival of Molly Southbourne*. 2019: Tom Doherty.

Thyfault, J. P., et al. *Physiology of sedentary behavior and its relationship to health outcomes*. Medicine and Science in Sports and Exercise, 2015. **47**(6): 1301.

Timothy, A., et al. *Influence of early adversity on cortisol reactivity, SLC6A4 methylation and externalizing behavior in children of alcoholics*. Progress in Neuro-Psychopharmacology and Biological Psychiatry, 2019. **94**:109649.

Tomiyama, A. J., et al. *How and why weight stigma drives the obesity "epidemic" and harms health*. BMC Medicine, 2018. **16**(1): 123.

Torondel, B., et al. *Association between unhygienic menstrual management practices and prevalence of lower reproductive tract infections: A hospital-based cross-sectional study in Odisha, India*. BMC Infectious Diseases, 2018. **18**(1): 1–12.

Towghi, F. *Haunting expectations of hospital births challenged by traditional midwives*. Medical Anthropology, 2018. **37**(8): 674–87.

Toze, M. *The risky womb and the unthinkability of the pregnant man: Addressing trans masculine hysterectomy*. Feminism and Psychology, 2018. **28**(2): 194–211.

Triebwasser, J. E., et al. *Pharmacy claims data versus patient self-report to measure contraceptive method continuation*. Contraception, 2015. **92**(1): 26–30.

Trussell, J. *Menarche and fatness: Reexamination of the critical body composition hypothesis*. Science, 1978. **200**:1506–9.

Trussell, J. *Statistical flaws in evidence for the Frisch hypothesis that fatness triggers menarche*. Human Biology, 1980. **52**(4): 711.

Tschudin, S., P. C. Bertea, and E. Zemp. *Prevalence and predictors of premenstrual syndrome and premenstrual dysphoric disorder in a population-based sample*. Archives of Women's Mental Health, 2010. **13**(6): 485–94.

Tsing, A. L. *The Mushroom at the End of the World: On the Possibility of Life in Capitalist Ruins*. 2015: Princeton University Press.

Tu, F. F., et al. *The influence of prior oral contraceptive use on risk of endometriosis is conditional on parity*. Fertility and Sterility, 2014. **101**(6): 1697–704.

Urquia, M., et al. *Disparities in pre-eclampsia and eclampsia among immigrant women giving birth in six industrialised countries*. BJOG: An International Journal of Obstetrics and Gynaecology, 2014. **121**(12): 1492–500.

Ussher, J. M. *The role of premenstrual dysphoric disorder in the subjectification of women*. Journal of Medical Humanities, 2003. **24**(1): 131–46.

Ussher, J. M., and J. Perz. *Empathy, egalitarianism and emotion work in the relational negotiation of PMS: The experience of women in lesbian relationships*. Feminism and Psychology, 2008. **18**(1): 87–111.

Vanden Brink, H., et al. *Age-related changes in major ovarian follicular wave dynamics during the human menstrual cycle*. Menopause, 2013. **20**(12): 1243–54.

Van der Molen, R., et al. *Menstrual blood closely resembles the uterine immune micro-environment and is clearly distinct from peripheral blood*. Human Reproduction, 2014. **29**(2): 303–14.

Vercellini, P., et al. *Association between endometriosis stage, lesion type, patient characteristics and severity of pelvic pain symptoms: A multivariate analysis of over 1000 patients*. Human Reproduction, 2007. **22**(1): 266–71.

Vercellini, P., et al. *Oral contraceptives and risk of endometriosis: A systematic review and meta-analysis.* Human Reproduction Update, 2010. **17**(2): 159–70.

Verma, S., P. E. Szmitko, and E.T.H. Yeh. *C-reactive protein: Structure affects function.* Circulation, 2004. **109**(16): 1914–17.

Villareal, A. *"No Data" Linking Covid Vaccines to Menstrual Changes, US Experts Say.* Guardian (United States), April 23, 2021.

Vitzthum, V. J., et al. *A prospective study of early pregnancy loss in humans.* Fertility and Sterility, 2006. **86**(2): 373–79.

Vitzthum, V. J., et al. *Vaginal bleeding patterns among rural highland Bolivian women: Relationship to fecundity and fetal loss.* Contraception, 2001. **64**:319–25.

Vostral, S. L. *Under Wraps: A History of Menstrual Hygiene Technology.* 2008: Lexington Books.

Walberg-Rankin, J., W. Franke, and F. Gwazdauskas. *Response of beta-endorphin and estradiol to resistance exercise in females during energy balance and energy restriction.* International Journal of Sports Medicine, 1992. **13**(7): 542–47.

Walker, R., et al. *Growth rates and life histories in twenty-two small-scale societies.* American Journal of Human Biology, 2006. **18**(3): 295–311.

Wamoyi, J., et al. *Transactional sex amongst young people in rural northern Tanzania: An ethnography of young women's motivations and negotiation.* Reproductive Health, 2010. **7**(1): 2.

Wander, K., E. Brindle, and K. A. O'Connor. *C-reactive protein across the menstrual cycle.* American Journal of Physical Anthropology, 2008. **136**(2): 138–46.

Wang, H., and S. K. Dey. *Roadmap to embryo implantation: Clues from mouse models.* Nature Reviews Genetics, 2006. **7**(3): 185–99.

Washington, H. A. *Medical Apartheid: The Dark History of Medical Experimentation on Black Americans from Colonial Times to the Present.* 2006: Doubleday.

Watson, R. E., T. B. Nelson, and A. L. Hsu. *Fertility considerations: The COVID-19 disease may have a more negative impact than the COVID-19 vaccine, especially among men.* Fertility and Sterility, March 19, 2021. https://www.fertstertdialog.com/posts/fertility-considerations-the-covid-19-disease-may-have-a-more-negative-impact-than-the-covid-19-vaccine-especially-among-men?room_id=871-covid-19.

Webster, V., P. Brough, and K. Daly. *Fight, flight or freeze: Common responses for follower coping with toxic leadership.* Stress and Health, 2016. **32**(4): 346–54.

Weinand, J. D., and J. D. Safer. *Hormone therapy in transgender adults is safe with provider supervision: A review of hormone therapy sequelae for transgender individuals.* Journal of Clinical and Translational Endocrinology, 2015. **2**(2): 55–60.

Werner, A., L. W. Isaksen, and K. Malterud. *"I am not the kind of woman who complains of everything": Illness stories on self and shame in women with chronic pain.* Social Science and Medicine, 2004. **59**(5): 1035–45.

White, L. R. *The function of ethnicity, income level, and menstrual taboos in postmenarcheal adolescents' understanding of menarche and menstruation.* Sex Roles, 2013. **68**(1): 65–76.

White, S., et al. *All the ACEs: A chaotic concept for family policy and decision-making?* Social Policy and Society, 2019. **18**(3): 457–66.

White Junod, S., and L. Marks. *Women's trials: The approval of the first oral contraceptive pill in the United States and Great Britain.* Journal of the History of Medicine, 2002. **57**:117–60.

Wilcox, A. J., D. D. Baird, and C. R. Weinberg. *Time of implantation of the conceptus and loss of pregnancy.* New England Journal of Medicine, 1999. **340**(23): 1796–99.

Williams, C. *"No Concerns That We're Seeing": Utah Doctors Explain How COVID-19 Vaccines Factor in Women's Health, Pregnancy.* KSL.com, April 16, 2021.

Williams, N. I., S. M. Statuta, and A. Austin. *Female athlete triad: Future directions for energy availability and eating disorder research and practice.* Clinics in Sports Medicine, 2017. **36**(4): 671–86.

Willis, C. *Daisy in the sun.* In *Fire Watch: Memorable Tales by a Young Master,* ed. C. Willis. 1985: Bluejay Books.

Willis, C. *Even the queen.* In *Impossible Things,* ed. C. Willis. 1994: Bantam Spectra.

Willness, C. R., P. Steel, and K. Lee. *A meta-analysis of the antecedents and consequences of workplace sexual harassment.* Personnel Psychology, 2007. **60**(1): 127–62.

Wilson, K., B. Sonenstein, and M. Kaba. *Hope Is a Discipline.* Interview by Kim Wilson and Brian Sonenstein, Beyond Prisons, January 2018. https://www.beyond-prisons.com/home/hope -is-a-discipline-feat-mariame-kaba.

Wister, J. A., M. L. Stubbs, and C. Shipman. *Mentioning menstruation: A stereotype threat that diminishes cognition?* Sex Roles, 2013. **68**(1): 19–31.

Wojtys, E. M., et al. *Athletic activity and hormone concentrations in high school female athletes.* Journal of Athletic Training, 2015. **50**(2): 185–92.

Wrangham, R. W. *Evolution of coalitionary killing.* Yearbook of Physical Anthropology, 1999. **42**:1–30.

Wrangham, R. W., and D. Peterson. *Demonic Males: Apes and the Origins of Human Violence.* 1996: Houghton Mifflin Harcourt.

Wunder, D. M., et al. *Serum leptin and C-reactive protein levels in the physiological spontaneous menstrual cycle in reproductive age women.* European Journal of Endocrinology, 2006. **155**(1): 137–42.

Yaszek, L. *The women history doesn't see: Recovering midcentury women's SF as a literature of social critique.* Extrapolation, 2004. **45**(1): 34–51.

Yeung, E. H., et al. *Adiposity and sex hormones across the menstrual cycle: The BioCycle study.* International Journal of Obesity, 2013. **37**(2): 237.

Zhu, Z., et al. *Mortality and morbidity of infants born extremely preterm at tertiary medical centers in China from 2010 to 2019.* JAMA Network Open, 2021. **4**(5): e219382.

Ziomkiewicz, A., et al. *Body fat, energy balance and estradiol levels: A study based on hormonal profiles from complete menstrual cycles.* Human Reproduction, 2008. **23**(11): 2555–63.

ILLUSTRATION CREDITS

Chapter One

Page 34: Macht, D. I., and D. S. Lubin, *A phyto-pharmacological study of menstrual toxin*, Journal of Pharmacology and Experimental Therapeutics, 1923. 22(5): 413–66.

Chapter Two

Page 53: (Left) Courtesy of Warren Anatomical Museum collection, Center for the History of Medicine in the Francis A. Countway Library of Medicine, Harvard University. Photograph by Samantha van Gerbig.

Page 53: (Right) Cleveland Museum of Natural History.

Page 78: Courtesy of the author.

Page 80: Courtesy of the author.

Chapter Three

Page 94: © Bettman/Getty.

Page 96: Redrawn from Ellison (1990); courtesy of the author.

INDEX

DISCUSSION QUESTIONS

1. In the introduction to *Period*, Clancy defines terms like "women" and "people who menstruate" and addresses why language specificity matters both for inclusion and for science. Does this language change the way you think about terms you use to describe people? Is this a book for people who have periods, or for women? What might non-menstruating people stand to gain from reading this book?

2. Clancy explains that when starting to write *Period*, she realized couldn't write "a straight science book," due to the long history of eugenics in the field of anthropology and gynecology and how much this is felt in reproductive justice today. Is it ever possible to disentangle bodies from culture?

3. Clancy begins the book by discussing some of what she learned about periods in her youth and how these lessons failed to capture the complexity and wonder of menstruation. How does this compare with your experience?

4. A pervasive misconception is that menstrual fluid is useless waste. But Clancy notes that "biomedical engineers are even starting to use the study of the endometrium and menstrual effluent to inform research and treatment for other conditions, ranging from wound repair to cardiac health." Could a wider understanding of this fact change the way periods are appreciated?

5. Clancy writes that "if we truly want to understand menstruation, it's better to seek to understand processes rather than outcomes and to explore variation." Why is it so important to understand the process? How do we benefit from accepting variation and not focusing too closely on norms?

6. Throughout the book, Clancy makes note of where sexist notions of feminine passivity have wrongly shaped dominant narratives in medicine and science. Where is this most prevalent when it comes to periods? Has reading *Period*, and thinking about sexism in science, made you skeptical of how other biological processes are discussed?

7. In chapter four, Clancy writes about menstrual hygiene management projects. She emphasizes that this work has "undoubtedly a very worthy goal," while offering critiques of where it falls short. What would it look like to have communities generate their own questions and design their own interventions?

8. Clancy devotes a chapter to the impact of stress on one's menstrual cycle. What are some structural sources of stress? What actions and support systems can individuals—both people who menstruate and those who don't—take and put in place to mitigate psychosocial stressors for their friends, family, and community members?

9. In the final chapter, Clancy writes, "I'd love to consider a broader period future: one that also creates more room for menstruating bodies . . . More than self-care or body positivity, I am advocating for the radical (but not new or original) idea that humans deserve dignity and that dignity means not only accommodating but celebrating and noticing all people." What do you think a better period future might look like?

10. Clancy writes that if "all we ever learn is that menstrual cycles make us hormonal, irritable, bloated, angry, depressed, anxious, or in pain, is it any wonder that's the primary way many of us perceive our experiences?" Do you think about periods differently after reading *Period*? Will the book affect how you talk about menstruation?